Design and Applications of Heterogeneous Nanostructured Materials

Design and Applications of Heterogeneous Nanostructured Materials

Guest Editors

Hongbo Ju
Bingyang Ma
Manuel António Peralta Evaristo
Jicheng Ding
Filipe Fernandes

MDPI

Basel • Beijing • Wuhan • Barcelona • Belgrade • Novi Sad • Cluj • Manchester

Guest Editors
Hongbo Ju
School of Material Science and Engineering
Jiangsu University of Science and Technology
Zhenjiang
China

Bingyang Ma
School of Materials Science and Engineering
Shanghai Dianji University
Shanghai
China

Manuel António Peralta Evaristo
Department of Mechanical Engineering
University of Coimbra
Coimbra
Portugal

Jicheng Ding
School of Materials Science and Engineering
Anhui University of Technology
Maanshan
China

Filipe Fernandes
CIDEM, ISEP
Polytechnic of Porto
Porto
Portugal

Editorial Office
MDPI AG
Grosspeteranlage 5
4052 Basel, Switzerland

This is a reprint of the Special Issue, published open access by the journal *Nanomaterials* (ISSN 2079-4991), freely accessible at: https://www.mdpi.com/journal/nanomaterials/special_issues/26E1VVYM3Z.

For citation purposes, cite each article independently as indicated on the article page online and as indicated below:

Lastname, A.A.; Lastname, B.B. Article Title. *Journal Name* **Year**, *Volume Number*, Page Range.

ISBN 978-3-7258-5829-3 (Hbk)
ISBN 978-3-7258-5830-9 (PDF)
https://doi.org/10.3390/books978-3-7258-5830-9

Contents

nanomaterials

MDPI

Article

High-Temperature Oxidation and Wear Resistance of TiAlSiN/AlCrN Multilayer Coatings Prepared by Multi-Arc Ion Plating

Jie Liu [1], Haijuan Mei [2,*], Junfang Hua [1], Juan Wang [3], Yongchao Wang [1], Genmiao Yi [1] and Xin Deng [4,*]

1 School of Intelligent Manufacturing, Guangzhou Panyu Polytechnic, Guangzhou 511483, China; liujie@gzpyp.edu.cn (J.L.); huajf@gzpyp.edu.cn (J.H.); wangyongc@gzpyp.edu.cn (Y.W.); yigm@gzpyp.edu.cn (G.Y.)
2 Guangdong Provincial Key Laboratory of Electronic Functional Materials and Devices, Huizhou University, Huizhou 516007, China
3 School of Art and Design, Guangdong University of Technology, Guangzhou 510006, China; wangjuan@gdut.edu.cn
4 School of Electromechanical Engineering, Guangdong University of Technology, Guangzhou 510006, China
* Correspondence: haijuanmei@hzu.edu.cn (H.M.); dengxin@gdut.edu.cn (X.D.)

Abstract: TiAlSiN and AlCrN coatings are two representative coatings with excellent properties in TiN-based and CrN-based coatings, respectively. Multilayering is one of the most important directions for coating performance optimization. In this paper, nanoscale monolayer TiAlSiN, AlCrN, and multilayer TiAlSiN/AlCrN coatings were prepared. The microstructure, mechanical properties, oxidation resistance, and wear resistance of the above three coatings were investigated. The following properties of the TiAlSiN/AlCrN coating, including phase, nanohardness, elastic modulus, adhesion strength, and oxidation resistance, fall between those of the TiAlSiN and AlCrN coatings and conform to the "law of mixtures". Due to the interfacial effect of the multilayer coating, the residual stress of the TiAlSiN/AlCrN coating is less than that of the two monolayer coatings. At 500 °C, the order of wear resistance of the three coatings is consistent with the order of H^3/E^{*2} values, i.e., TiAlSiN > TiAlSiN/AlCrN > AlCrN; at 800 °C, the order of wear resistance becomes TiAlSiN/AlCrN > TiAlSiN > AlCrN because TiAlSiN coating has entered the rapid oxidization stage first, reducing its wear resistance.

Keywords: multilayer; TiAlSiN; AlCrN; oxidation resistance; wear; mechanical properties

1. Introduction

Among Ti-based and Cr-based coatings, TiN and CrN are the two earliest and most widely used. Their performance can be further improved by adding other elements. Al and Si are the two most successful additive elements. Al atoms strengthen solid solutions by replacing Ti in the lattice of TiN, but too much Al will transform TiAlN from a single FCC phase to a mixed phase structure of FCC and HCP, decreasing the coating's mechanical properties. In addition, the dense and chemically stable Al_2O_3 generated by the oxidation of Al is also responsible for the improved resistance to oxidation of Al-containing coatings. AlTiN- and AlCrN-coated tools have been successfully used in machining because of their excellent hardness and high-temperature oxidation resistance [1,2]. Since TiO_2 undergoes a phase transition from dense α-TiO_2 to loose r-TiO_2 at high temperatures, whereas Cr_2O_3 does not have such a change, AlCrN has superior resistance to oxidation than AlTiN.

Si was added to TiAlN to obtain TiAlSiN coating. Nanocomposite structures have been proven to exist in such coating [3,4]. This structure consists of amorphous Si_3N_4

surrounding TiN grains, while Al is solidly dissolved in the TiN lattice. This nanocomposite structure hinders dislocations and refines the grains, thus enhancing the coating's hardness [5], which will contribute to its wear resistance. Coating wear resistance is also related to the modulus of elasticity E. Many researchers have used the H/E^* and H^3/E^{*2} values to comprehensively evaluate the effects of hardness and modulus of elasticity on coating wear resistance [6,7]. The nanocomposite structure can inhibit the TiN grains encapsulated by the amorphous phase from growing up at high temperatures and also prevent the intrusion of oxygen, so TiAlSiN obtains excellent high-temperature stability, as well as oxidation resistance [8,9]. This will help it be used in high-temperature, high-stress applications such as high-speed machining of difficult-to-cut materials [10,11].

In recent years, making coatings into multilayers has become an effective means of obtaining comprehensive coating performance. Due to their unique interfacial effects, multilayer coatings offer outstanding toughness, corrosion resistance, and thermal stability over monolayer coatings [12]. In the study of CrN/ZrN [13] multilayer coatings, the oxidation resistance of the coatings was found to increase as the number of layers increased. The modulation period has a decisive influence on multilayer coatings' microstructure and mechanical properties. In the study of TiAlSiN/CrN [14] multilayer coatings, it was found that the columnar crystal grain size increased with decreasing modulation period, while the hardness and elastic modulus first increased and then decreased. Adjusting the composition and thickness of the different layers within multilayer coatings can improve the coating's wear resistance. In the study of AlCrBN/AlTiBN [15] multilayer coatings, the lowest wear rate of $8.7 \pm 2.3 \times 10^{-7}$ mm^3/Nm was obtained when the thicknesses of AlTiBN and AlCrBN were set to 6.5 nm and 16.7 nm, respectively. Baijun Xiao [16], in his experiments on turning SKD11 (equivalent to AISI D2 steel), found that the AlCrN/AlTiSiN-coated tool had a longer life of about 800 m compared to the monolayer AlCrN and AlTiSiN coatings. This can be attributed to its higher hardness, adhesion strength, and resistance to oxidation.

Past studies have focused on the properties or preparation of TiAlSiN or AlCrN coatings. However, fewer comparative studies have been conducted between the two coatings and the multilayer coating made from both of them, and the conclusions have not been inconsistent.

2. Materials and Methods

2.1. Coating Deposition

The coatings in this paper were manufactured using the multi-arc ion plating technique (NH-10758 multifunctional PVD coater, Dongguan Nahu Crystal Materials Co., Dongguan, China). The coated substrate was fitted on a rotating shelf inside the furnace chamber and has three dimensions of rotational motion, as illustrated in Figure 1a. For monolayer coating, targets are loaded on one side only, either on the left or right; for multilayer coating, targets of different compositions are loaded on opposite sides. The structure and composition of TiAlSiN/AlCrN multilayer coating were designed according to Figure 1b. The modulation period of the multilayer coating is determined by the rotational speed of the rotating shelf. In our team's published study [17], the rotational speed of the rotating shelf was set to 2 r/min, and TiAlSiN/AlCrN multilayer coating with a 20 nm modulation period was obtained. This nanoscale modulation period was confirmed by TEM. In this paper, the rotational speed of the rotating shelf was doubled to 4 r/min, and a 10 nm modulation period of TiAlSiN/AlCrN multilayer coating will be obtained.

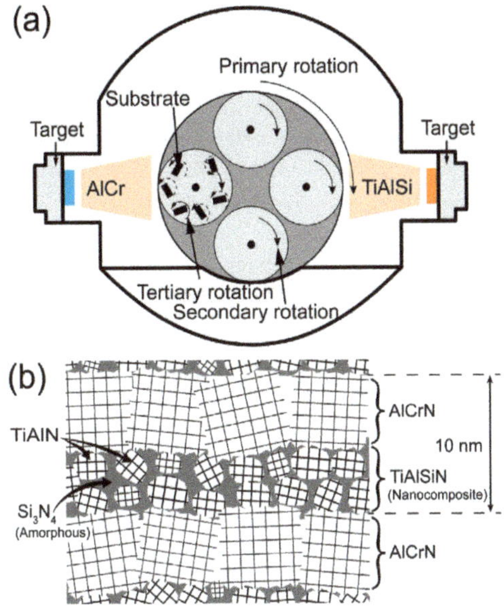

Figure 1. Illustration of coating fabrication and coating structure: (**a**) fabrication equipment for multilayer coating; (**b**) structure of multilayer coating.

The coated substrates are as follows: a WC cemented carbide block with a Co mass fraction of 6% and a size of 18 × 18 × 5 mm for mechanical property testing and fracture cross-section observation; single-crystal alumina with an outer diameter of 51 mm and a thickness of 0.5 mm for high-temperature oxidation test; a polycrystalline alumina with dimensions of 23 × 10 × 0.5 mm for thermogravimetric analysis (TGA); and an AISI 304 stainless steel sheet with dimensions of 60 × 10 × 0.8 mm for residual stress testing. The substrates were polished, cleaned, and put into the furnace chamber, which was evacuated to 5.0×10^{-3} Pa and heated to 500 °C. Next, the substrate was glow-cleaned for 9.5 min by passing 0.6 Pa argon gas at a bias voltage of 800 V and was then anodically etched for 20 min at a bias voltage of 300 V to further clean and activate the substrate surface. Then, nitrogen was passed in three stages at pressures of 2 Pa, 3 Pa, and 3 Pa, corresponding to bias voltages of 30 V, 40 V, and 60 V, and target currents of 145 A, 160 A, and 160 A, respectively. The above changes form a gradient layer, reducing the residual stress and enhancing the adhesion between the coating and the substrate. To achieve the same thickness of 3.5 μm for all three coatings, the deposition times of TiAlSiN, AlCrN, and TiAlSiN/AlCrN coatings were 194 min, 111 min, and 74 min, respectively. Finally, the samples were taken out after the temperature was lowered to room temperature.

2.2. Coating Characterization

The surface and fracture cross-section morphology and composition of the coatings were measured using a Nova NanoSEM 430 (FEI, Eindhoven, The Netherland). Phase analysis of the coatings was carried out using a Bruker D8 Advance X-Ray Diffractometer (Bruker Corp, Billerica, MA, USA). The parameters were set to a 2θ angle of 25°~85°, a step size of 0.02°, a dwell time of 0.5 s per step, and an incidence angle of 1° for grazing incidence XRD (GIXRD). The dwell time per step for conventional X-ray diffraction (XRD) was adjusted to 0.3 s. The coatings' nanohardness (H) and elastic modulus (E) were measured by nanoindentation using Anton Paar's TTX-NHT2 (Anton Paar, Graz, Austria). Test conditions were established with an applied load of 10 mN, a loading rate of 15 mN/min, a maximum depth of 120 nm, and a load retention time of 5 s at the maximum depth. It was necessary to restrict the indentation depth to 10% of the coating thickness in order to minimize substrate interference with the coating's nanohardness. The adhesion strength between the coating and the substrate was measured using an Anton Paar RST3 Scratch-meter (Anton Paar, Graz, Austria). The parameters were set as a diamond tip diameter of 200 μm, an applied load ranging from 1 to 100 N with a loading rate of 200 N/min, and a scratch length of 3 mm. The residual stresses in the coating were measured using the Supro Instruments FST-1000 Film Stress Tester (SuPro Instruments Ltd., Shengzhen, China) and then calculated using the Stoney equation [18]. TGA testing of the coatings was carried out using a NETZSCH STA449F5 thermogravimetric analyzer (NETZSCH Group, Bavarian, Germany). The heating rate was 10 K/min, and the dry airflow was maintained at 50 sccm during the entire experiment. Oxidation experiments on the coatings were carried out in a TSX1700 muffle furnace (Cinite, Beijing, China). The coatings were exposed to temperatures of 800 °C, 950 °C, and 1050 °C with a 3 h dwell time.

The coatings' resistance to wear was tested using an Anton Paar THT high-temperature ball and disk friction tester (Anton Paar, Graz, Austria). The counterbody was a polycrystalline Al_2O_3 ball, measuring 6 mm in diameter. The test parameters were load 10 N, linear velocity 0.1 m/s, friction radius 2 mm, and the number of friction circles 8000 at 500 °C and 6000 at 800 °C, equivalent to a distance of 100 m and 75 m, respectively. A lower number of cycles is set at higher temperatures to prevent the coating from being worn through at elevated temperatures.

The formula for calculating the coating wear rate is as follows:

$$W = \frac{V}{L \times P} \tag{1}$$

where W is the wear rate, V is the volume of material lost in the friction experiment in mm^3, L is the total length of the friction in m, and P is the load in N. V is equal to the product of the cross-sectional area of the wear track and the circumference of the circle, the cross-sectional area of the wear track is measured in the laser confocal software, and L is calculated as the product of the number of circles and the circumference.

Surface roughness tends to make the nanoindentation data discrete. Five randomly selected test points on the surface were needed, and the data obtained were evaluated for variability by calculating the mean and standard deviation. Outliers were identified using the Grubbs test and excluded if they were significantly out of the expected range. Scratch, residual stress, and cross-sectional area of the wear track were measured three times each to ensure data reliability and repeatability.

3. Results and Discussion

3.1. Phase and Microstructure

Figure 2 presents the fracture cross-sections and surface morphologies of the TiAlSiN, AlCrN, and TiAlSiN/AlCrN coatings. The AlCrN coating exhibits distinct columnar crystal structures, whereas the TiAlSiN and TiAlSiN/AlCrN coatings display featureless morphologies, typical of Si-containing coatings. Numerous white microparticles are evenly distributed across the surface of the coatings, which exhibits a typical deposition pattern of arc ion plating. This is because the cathodic arc produces an arc spot on the target's surface with a small breakdown area and, thus, a high-power density. Under the combined effect of internal expansion pressure and negative bias pressure at the substrate end, the metal in the molten pool splashes onto the substrate surface and forms particles before it can evaporate [19,20]. Among these target materials, Al has the lowest melting point.

Additionally, the AlCrN coating contains the highest Al content among the three coatings. Consequently, the AlCrN coating exhibits the most white microparticles on its surface, followed by the TiAlSiN/AlCrN coating.

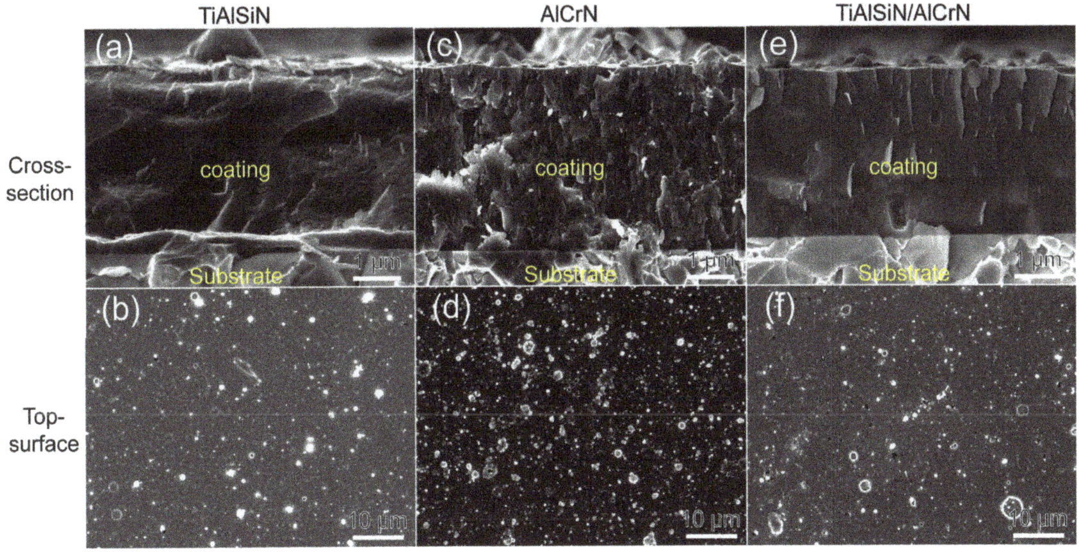

Figure 2. Morphology of fracture cross-section and top surface for three coatings: (**a**,**b**) TiAlSiN; (**c**,**d**) AlCrN; (**e**,**f**) TiAlSiN/AlCrN.

The chemical compositions of the three coatings were analyzed using EDS, and the results are presented in Table 1. The elemental ratios closely match the target. However, the Si content is lower than the target due to its lower deposition rate compared to Al and Ti.

Table 1. Coating thickness and composition measured by SEM and EDS.

Target	Coating	Thickness (μm)	Chemical Composition (at.%)				
			Al	Cr	Ti	Si	N
$Ti_{0.45}Al_{0.45}Si_{0.10}$	TiAlSiN	3.69	20.22	–	22.66	3.89	53.23
$Al_{0.7}Cr_{0.3}$	AlCrN	3.63	33.81	18.14	–	–	48.05
$Ti_{0.45}Al_{0.45}Si_{0.10}$ and $Al_{0.7}Cr_{0.3}$	TiAlSiN/AlCrN	3.52	27.43	7.38	11.71	2.27	51.21

Figure 3 shows the phase analysis results of the coatings obtained using GIXRD. According to ICDD No. 38-1420 for c-TiN and ICDD No. 11-0065 for c-CrN, the diffraction peaks corresponding to the (111), (200), (220), (311), and (222) planes of the cubic phases of c-TiAlN and c-CrAlN are observed in both TiAlSiN and AlCrN coatings. The TiAlSiN coating exhibits a pronounced TiN (200) plane preferred orientation, whereas the AlCrN has a CrN (111) preferred orientation. The diffraction peaks of TiAlSiN/AlCrN coating appear between the standard peaks of c-TiN and c-CrN, indicating that the TiAlSiN/AlCrN coating exhibits a mixed cubic phase structure of c-TiAlCrN, characterized by a preferred orientation along the (200) plane. Distinct w-AlN phases (ICDD No. 25-1133) are observed in both the TiAlSiN and TiAlSiN/AlCrN coatings but not in the AlCrN coating, despite the higher Al content (70 at.%) present in the AlCrN coating. Other researchers have made similar findings [21]. This is attributed to the fact that the solubility of Al in TiN is less than that of CrN, and the introduction of Si further reduces the solubility of Al in TiN. Both factors result in the formation of w-AlN phases in the TiAlSiN and TiAlSiN/AlCrN coatings.

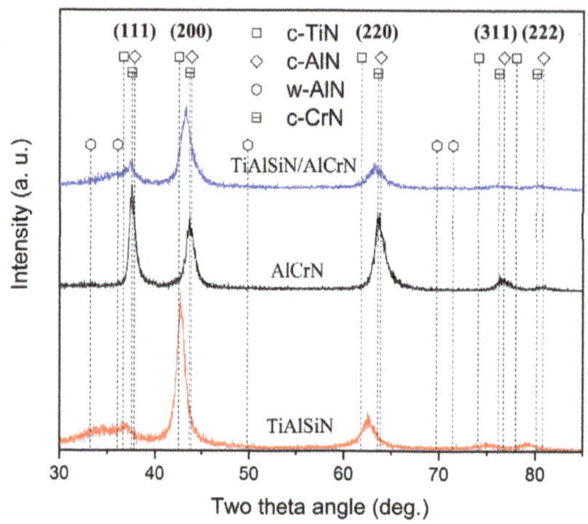

Figure 3. XRD patterns of three coatings.

3.2. Mechanical Properties

The nanohardness and elastic modulus of TiAlSiN, AlCrN, and TiAlSiN/AlCrN coatings are presented in Figure 4. The highest nanohardness and lowest elastic modulus are obtained for the TiAlSiN coating, while these two metrics for the TiAlSiN/AlCrN coating are in between those for the TiAlSiN and AlCrN monolayer coatings, which is by the "law of mixtures". A coating's plastic deformation and wear resistance can also be usually measured by H^3/E^{*2} [22,23], where E* is equal to $E(1-\mu^2)$. The higher the value of H^3/E^{*2}, the higher the coating's plastic deformation and wear resistance. Since TiAlSiN coating has the highest nanohardness and lowest modulus of elasticity, it has the highest H^3/E^{*2} value.

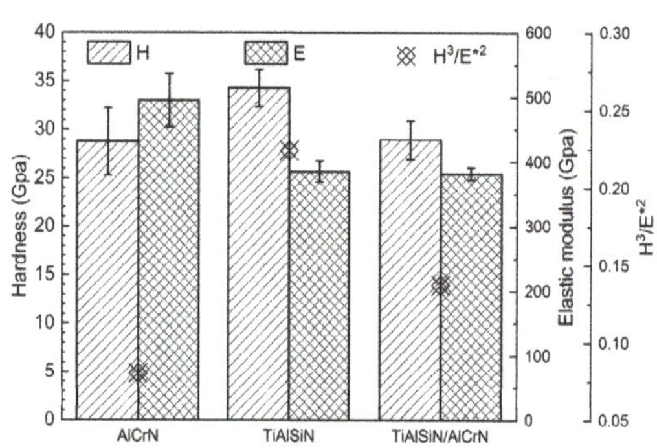

Figure 4. Nanohardness and elastic modulus of three coatings.

The adhesion strength between the coating and the substrate is one of the most important factors in evaluating a coating's performance. Figure 5 presents the critical Lc2 and Lc3 values of the three coatings measured during the scratch test. Lc2 represents the load at which cracks initiate and minor coating flaking occurs, whereas Lc3 indicates the load at which the substrate becomes exposed. Typically, Lc2 is regarded as the critical load for coating failure and serves as a measure of coating adhesion strength. The AlCrN coating exhibits the highest adhesion strength. In comparison, the TiAlSiN coating exhibits an Lc2 value of only 41 N, whereas the adhesion strength of the TiAlSiN/AlCrN coating falls between the other two coatings. The significant difference between the Lc2 and Lc3 values for the multilayer coating, along with the presence of powdery debris distributed on both sides of the scratches, indicates that the multilayer coating tends to gradually detach from the substrate in thin layers. Conversely, the slight difference between the Lc2 and Lc3 values of the two monolayer coatings indicates that these coatings are completely removed quickly upon reaching the critical load (Lc2). This rapid detachment is confirmed by the presence of large debris fragments on both sides of the scratches. After the addition of Si, the poor bonding of amorphous Si_3N_4 to the substrate crystals and the higher residual stresses within the coating result in reduced adhesion strength of the TiAlSiN coating [23]. The residual stress values of the three coatings are −8.3 GPa for TiAlSiN, −5.8 GPa for AlCrN, and −3.6 GPa for TiAlSiN/AlCrN, respectively. The lower stresses in the multilayer coating are attributed to the interfacial effect of the increased number of coating interfaces, which effectively release the internal stresses throughout the coating.

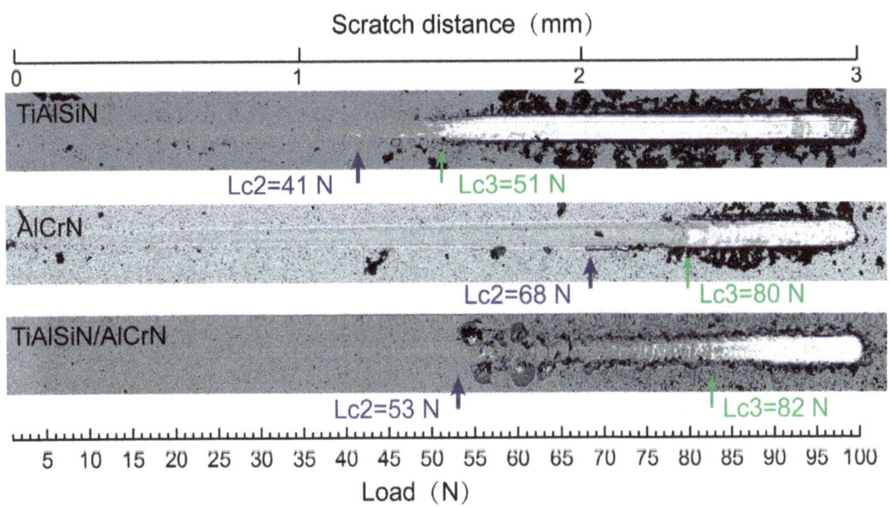

Figure 5. The critical Lc2 value and corresponding optical graphs after the scratch test of three coatings.

3.3. High-Temperature Oxidation Resistance

Figure 6 depicts the TGA curves of TiAlSiN, AlCrN, and TiAlSiN/AlCrN coatings subjected to continuous heating from room temperature to 1450 °C in synthetic air. The oxidized weight gain of the three coatings is imperceptible until 900 °C. From 950 °C to 1050 °C, the TiAlSiN coating's TGA curve suddenly forms a local peak upward, indicating the emergence of some kind of new oxidation mechanism. Further, the TiAlSiN coating enters a rapid oxidation stage. As the temperature increases to 1375 °C, the oxidation weight gain of the TiAlSiN coating remains unchanged, signifying that the coating has undergone complete oxidation and reached the terminated oxidation stage. On the contrary, the oxidation of the AlCrN coating is not complete by the end of the experiment at 1450 °C, suggesting that the oxidation resistance of the AlCrN coating exceeds that of TiAlSiN. Meanwhile, the initial and termination oxidation temperatures of the TiAlSiN/AlCrN multilayer coating are 1200 °C and 1400 °C, respectively, between the two monolayer coatings.

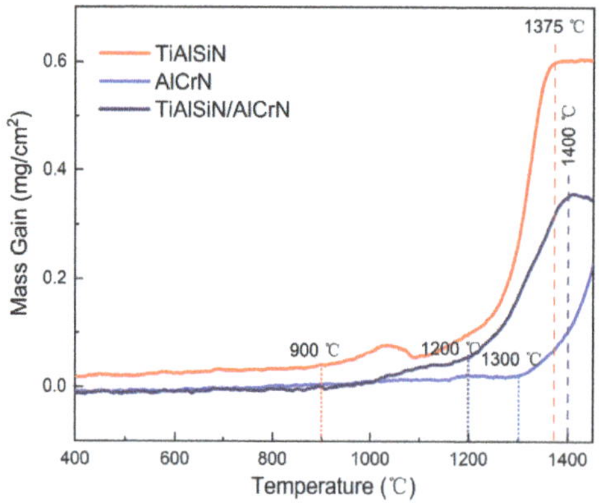

Figure 6. TGA curves of three coatings in synthetic air atmosphere.

For the oxidation details of the coatings, the three coatings were oxidized at 800 °C, 950 °C, and 1050 °C for three hours. Figure 7 shows the surface morphology after oxidation. From 800 °C to 1050 °C, the surface of the TiAlSiN coating undergoes the greatest change, with the number of oxides increasing significantly and the oxide particles gradually becoming larger. While the surface of the AlCrN coating has the smallest change, the magnitude of change in the TiAlSiN/AlCrN coating is between the above two.

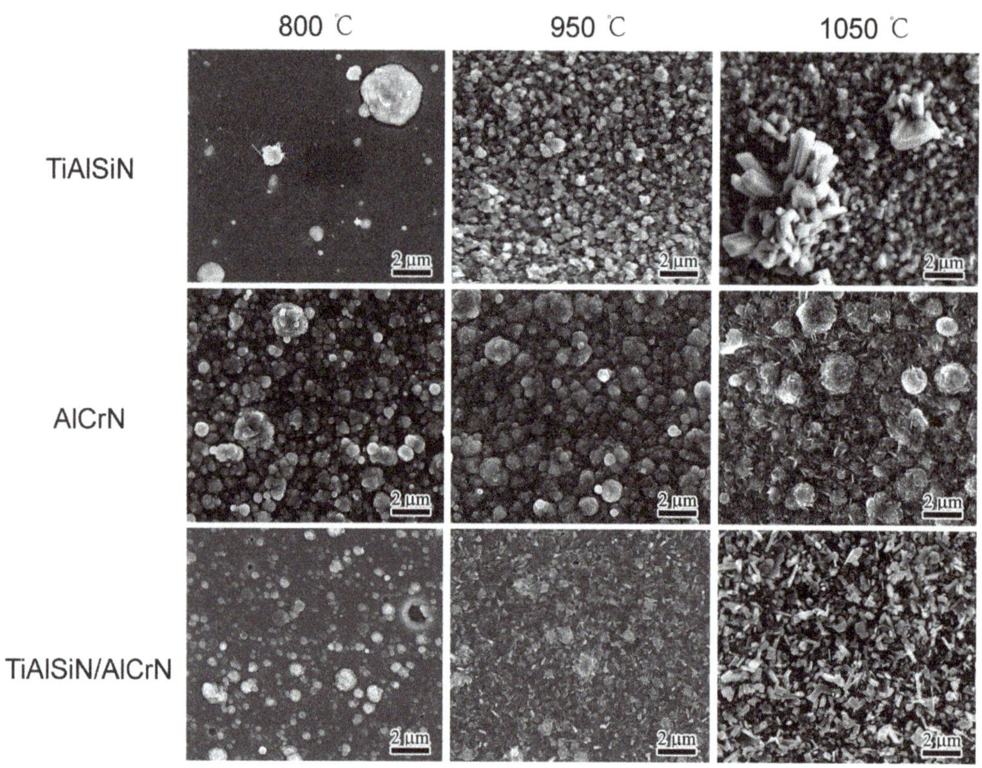

Figure 7. Morphology of top surface for three coatings after oxidizing at high temperature for three hours.

XRD was performed on the oxidized surfaces to identify the oxidation products, and the results are presented in Figure 8. At 800 °C, no oxide peaks appear on the surface of any of the three coatings. At 950 °C, three oxides, Al_2O_3 (ICDD No. 10-0173, ICDD No.26-0031, ICDD No. 46-1131), α-TiO_2 (ICDD No. 21-1272), and r-TiO_2 (ICDD No. 21-1276), appeared in the TiAlSiN coating, as shown in Figure 8a. At 1050 °C, the peak intensity of Al_2O_3 was enhanced, especially at a diffraction peak of 64.5°. The peak intensities of the two oxides of TiO_2 show opposite trends, with α-TiO_2 weakening, e.g., 25.2°, 48.0°, and 55.0°, and r-TiO_2 enhancing, e.g., 36.0° and 54.3°, suggesting that TiO_2 is in the process of phase transition from α-TiO_2 to r-TiO_2 from 950 °C to 1050 °C. The onset temperature of this phase transition is after 800 °C and before 950 °C. It is consistent with the findings of Kumar [24] and Shi [25]. They suggested that the phase transition from α-TiO_2 to r-TiO_2 occurs gradually between 800 °C and 1000 °C and is completed at 1100 °C to 1200 °C. The phase transition of TiO_2 explains the local peak near 1050 °C in the TGA curve of the TiAlSiN coating in Figure 6. From 950 °C to 1050 °C, the enhancement of the oxide peaks of both AlCrN and TiAlSiN/AlCrN coatings is almost negligible. The weaker intensity of their oxide peaks compared with TiAlSiN indicates that their oxidation resistance is significantly superior to that of TiAlSiN. The above results are consistent with the TGA results in Figure 6.

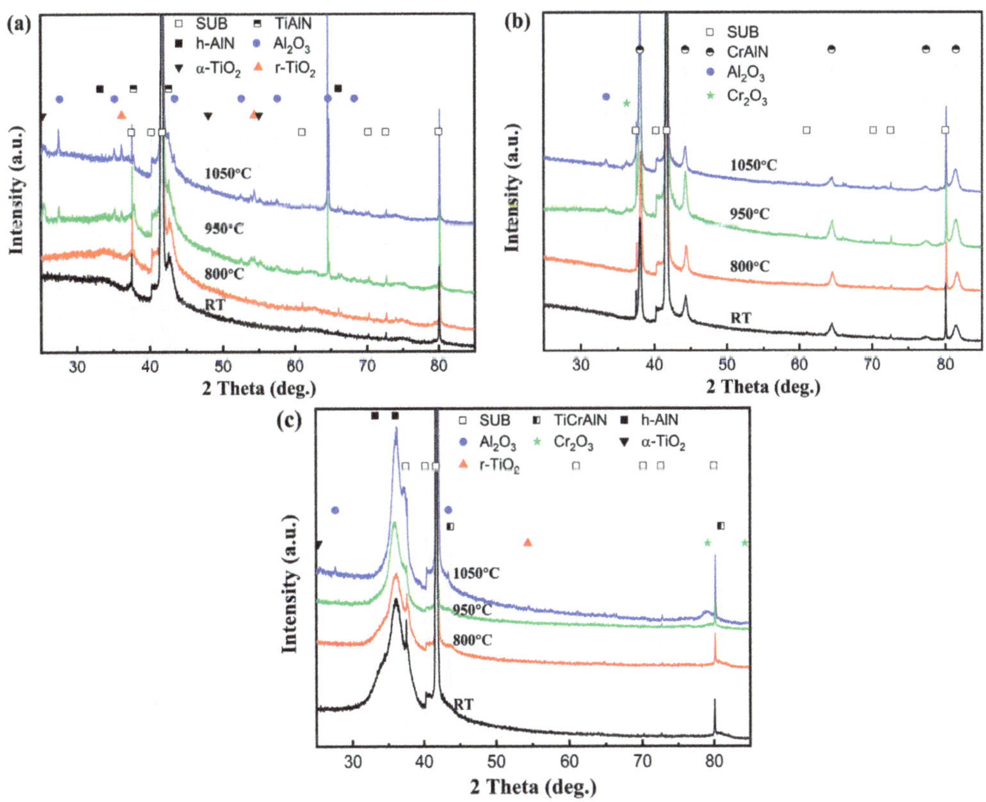

Figure 8. XRD patterns of coatings after high-temperature oxidation tests: (**a**) TiAlSiN; (**b**) AlCrN; (**c**) TiAlSiN/AlCrN.

SEM observations and EDS mapping were conducted on the fracture cross-sections of the three coatings oxidized at 950 °C and 1050 °C, respectively. At 950 °C, the oxides of the TiAlSiN coating showed a bilayer structure, as shown in Figure 9a,b. With reference to the XRD results, the upper layer is Al_2O_3, and the lower layer is TiO_2. Since $\Delta G^0 = -1336$ KJ/mol for Al_2O_3 and $\Delta G^0 = -753$ KJ/mol for TiO_2, Al_2O_3 is preferentially generated and located in the upper layer. The dense Al_2O_3 film covering the surface of the coating can block the diffusion of oxygen into the interior of the coating, thus improving the high-temperature oxidation resistance of the coating [26]. Since the thermal expansion coefficient of TiO_2 (10.5×10^{-6}/K) is larger than that of Al_2O_3 (8.4×10^{-6}/K) [27], when the temperature continues to increase to 1050 °C, the upper layer of Al_2O_3 is tensile cracked. The unoxidized and highly concentrated Ti in the lower layer diffuses along the crack to the upper layer. Therefore, as seen in Figure 9d, the upper layer of the oxide is transformed into a Ti-rich oxide, and the lower layer is a mixed Al-Ti oxide.

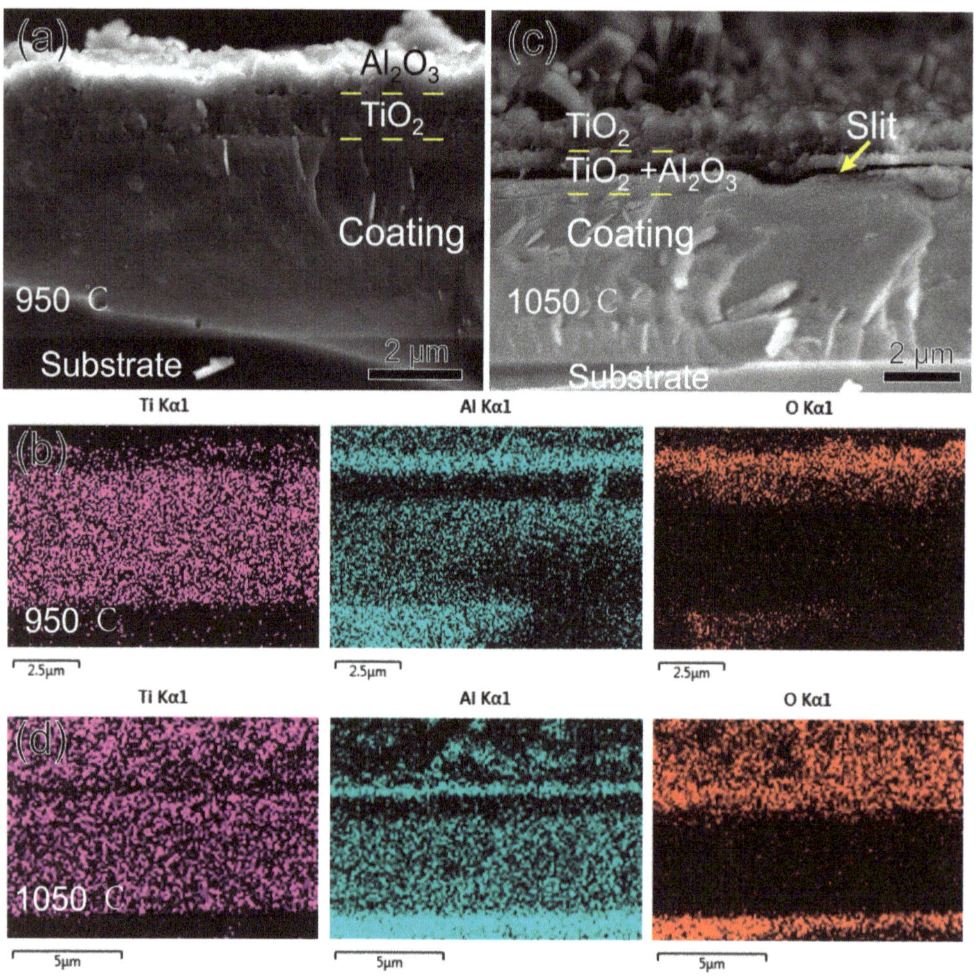

Figure 9. Morphology and EDS mapping of TiAlSiN coating fracture cross-section after oxidizing at high temperature for three hours: (**a**,**b**) 950 °C; (**c**,**d**) 1050 °C.

Compared to α-TiO_2, r-TiO_2 is loose and porous, so when it appears in large quantities, it is accompanied by the appearance of a large number of pores within the oxide, which allows for an additional mode of diffusion of oxygen and metal by mass transport in addition to the original short-circuit diffusion. The whole oxide layer is filled with r-TiO_2 from phase transition at 1050 °C, providing a convenient channel for the outward migration of Ti. When a significant amount of Ti diffuses away from the oxide–nitride interface, micropores are formed, and the micropores are further connected to form a slit, as shown in Figure 9c. Even so, oxidation at 1050 °C for three hours is slow, and the coating's oxidized thickness is less than 15% of its total thickness. The outstanding oxidation resistance of the TiAlSiN coating is attributed to the presence of Si. First, the nanocomposite structure of the coating extends the oxygen diffusion path; second, Si has the effect of preventing the transformation of TiO_2 from dense α-TiO_2 to looser r-TiO_2, which retards the oxidation of the coating [28]; and third, SiO_2, an oxide of Si, also plays a protective role in the antioxidant property of the coating [29]. SiO_2 yields were very low and not detected in the EDS or XRD test.

Figure 10 shows the morphology and EDS mapping of the fracture cross-section of the AlCrN coating after oxidation for three hours at 950 °C and 1050 °C. The oxide layer of the coating at 950 °C is also double-layered, as shown in Figure 10a,b. From the corresponding Al_2O_3 and Cr_2O_3 at 33.3° and 36.2°, respectively, in Figure 8b, it can be seen that the upper layer of the oxide is Al_2O_3, and the lower layer is a mixture of Al_2O_3 and Cr_2O_3, as observed by Xiang D. Zhang [30]. Al_2O_3 is located in the upper layer for the same reason because its ΔG^0 is much lower than that of Cr_2O_3, which is −845 KJ/mol, so Al_2O_3 is preferentially generated and located in the upper layer. At 1050 °C, the delamination at 950 °C is no longer present, and the whole oxide layer is a mixture of Al_2O_3 and Cr_2O_3. AlCrN coating did not replace the upper oxide layer at 1050°C like TiAlSiN coating because the thermal expansion coefficient of Cr_2O_3 is only 7×10^{-6}/K, less than that of Al_2O_3, and the thermal stresses could not tear apart the upper Al_2O_3 and release the Cr ions. However, at this temperature, the diffusion rate of Cr ions inside the oxide dramatically increases compared to Al [31,32]. The Cr gathered in the lower layer diffuses upwards and mixes with Al. We already know that Al_2O_3 and Cr_2O_3 have the same hexagonally symmetric corundum crystal structure, and the final formation of Cr_2O_3 and Al_2O_3 exists as a solid solution of $(Cr, Al)_2O_3$. Assuming a higher temperature or a longer duration of the experiment, delamination of the oxide with Cr_2O_3 located in the upper layer will occur, as seen by Yuxiang Xu [32] after oxidizing AlCrN at 1100 °C for 20 h.

Figure 11 shows the morphology and EDS-mapping of the fracture cross section of the TiAlSiN/AlCrN multilayer coating after oxidation for three hours at 950 °C and 1050 °C. At 950 °C, just as it is difficult to perceive the presence of the oxides in the XRD of Figure 8c, the thickness of the oxides shown in Figure 11a,b is significantly smaller than that of the other two monolayer coatings at this temperature. In addition, Al, Cr, or Ti oxides are not delaminated but mixed. This should be attributed to the numerous interlayer interfaces possessed by the multilayer coating, which act as barriers to oxygen and metal ion diffusion [33]. All metal ions, including Al, which has the strongest affinity for oxygen, are imprisoned in situ and oxidized. Therefore, the multilayer coating shows reduced oxidation and no delamination of the oxides at 950 °C. At 1050 °C, the oxide's thickness increased, and the Ti concentration in it enhanced significantly, indicating that the TiO_2 increased significantly. In addition, the same slit as the TiAlSiN coating appeared at the interface between the oxide and nitride, as shown in Figure 11c. All these phenomena are still caused by the r-TiO_2 coming from the phase transition at this temperature. The r-TiO_2 accompanying the pores weakens the multilayer structure's hindering effect on diffusion, resulting in increased oxidation and a thicker oxide layer. In addition, the involvement of r-TiO_2 makes it impossible to form a continuous dense (Cr, Al)$_2O_3$ film like AlCrN in the outermost layer of the oxide, so the surface of the oxide layer is rugged, as seen in Figure 11d. In this case, the barrier effect of the oxide layer decreases significantly. It has been proposed that Ti has a negative effect on the oxidation resistance of the AlCrN coating, attributed to the greater affinity of Ti for oxygen and the cracking of the (Cr, Al)$_2O_3$ surface film induced by the growth stress of TiO_2 [32]. However, in terms of affinity, Al has a greater affinity for O. In addition, the molar volume per unit oxygen atom of TiO_2 lies between Al_2O_3 and Cr_2O_3, so TiO_2 cannot be considered to have more significant growth stress. We argue that the more likely cause of TiO_2 damage to the (Cr, Al)$_2O_3$ surface film is the thermal stress determined by its maximum coefficient of thermal expansion, rather than its growth stress.

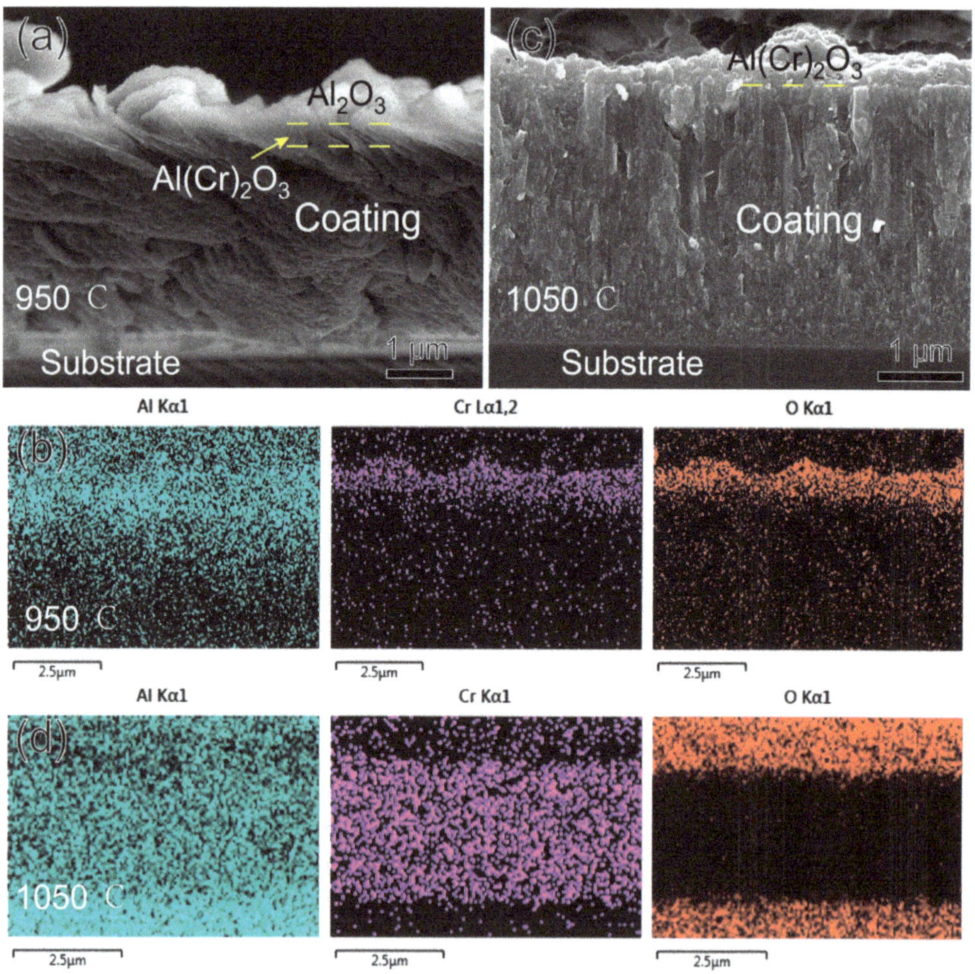

Figure 10. Morphology and EDS mapping of AlCrN coating fracture cross-section after oxidizing at high temperature for three hours: (**a**,**b**) 950 °C; (**c**,**d**) 1050 °C.

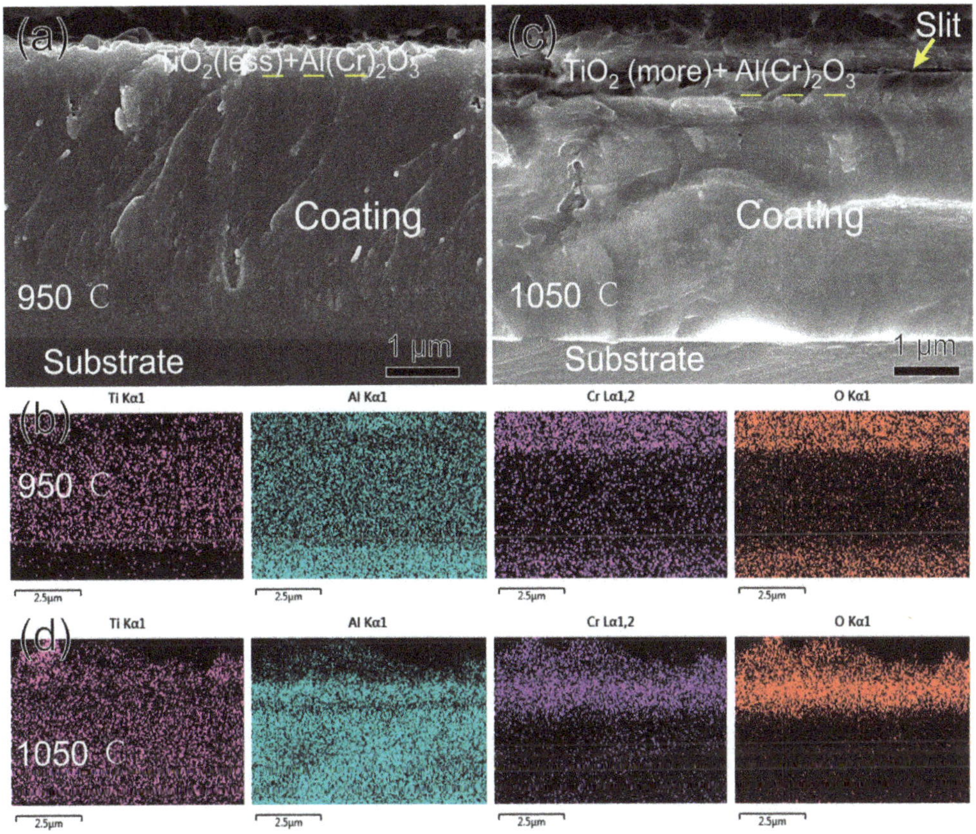

Figure 11. Morphology and EDS mapping of TiAlSiN/AlCrN coating fracture cross-section after oxidizing at high temperature for three hours: (**a**,**b**) 950 °C; (**c**,**d**) 1050 °C.

3.4. High-Temperature Wear Resistance

Figure 12 shows the friction coefficient curves of the three coatings, TiAlSiN, AlCrN, and TiAlSiN/AlCrN, after friction tests at two temperatures. First, the amplitudes of the curves at the two temperatures are analyzed. Compared to the ambient temperature of 800 °C, the curves of the three coatings at 500 °C fluctuate dramatically, especially for the AlCrN coating. This is caused by the original microscopic asperities and wear debris generated by friction on the contact surfaces. A large amount of wear debris is visible in SEM Figure 13f of the coating after the friction test. At 800 °C, the number of oxides involved in the friction process increases considerably. Although the XRD in Figure 8 does not show visible oxides, the friction surface temperature should be higher than the ambient temperature, so oxides exist. On one hand, these oxides fill in the grooves in the wear scar; on the other hand, they themselves act as a lubricant [8]. Second, the change in the mean friction coefficient from 500 °C to 800 °C is analyzed. The mean friction coefficient of the TiAlSiN/AlCrN coating decreases from 0.65 to 0.35 and that of the AlCrN coating decreases from 0.5 to 0.25, while that of the TiAlSiN coating shows no reduction. This is because the first two coatings both contain Cr_2O_3. Other researchers have also found that the lubricating effect of Cr_2O_3 reduces the friction coefficient [34]. The TiAlSiN coating achieves the lowest friction coefficient related to its smooth and hard surface. The TiAlSiN/AlCrN multilayer coating exhibits the highest friction coefficient, possibly due to the impact of the multilayer structure on the production of wear debris, requiring further investigation.

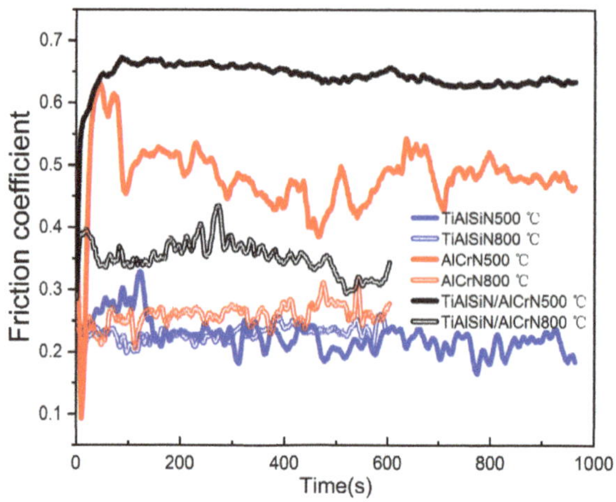

Figure 12. Friction coefficient curves of the three coatings at two temperatures.

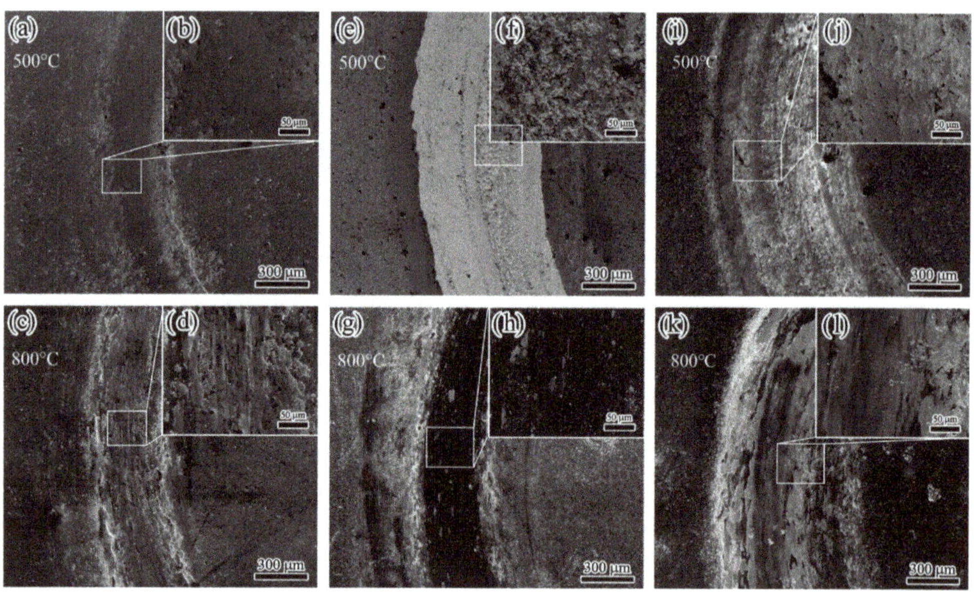

Figure 13. Wear tracks of three coatings after wear test at high temperatures: (**a–d**) TiAlSiN; (**e–h**) AlCrN; (**i–l**) TiAlSiN/AlCrN.

After the friction test, the morphology of the wear tracks of the three coatings is analyzed. At 500 °C, the AlCrN coating exhibited the most conspicuous wear tracks. The friction produced a large amount of debris, which accumulated on the wear tracks, as shown in Figure 13f. The TiAlSiN coating had imperceptible wear tracks; TiAlSiN/AlCrN was intermediate between the two. The wear rate histograms in Figure 14 support the above observations. The AlCrN coating has the highest wear rate, the TiAlSiN coating has the lowest, and the TiAlSiN/AlCrN coating is in the middle of the range. At 800 °C, compared to 500 °C, the wear tracks of the TiAlSiN coating became obvious, and a large amount of debris was generated, as shown in Figure 13d, indicating that the wear of this coating had worsened. In contrast, the wear tracks of the AlCrN and TiAlSiN/AlCrN coatings became smoother, suggesting that the wear of these two coatings was reduced. Again, the wear rate histograms support the above observation. The wear rate of the TiAlSiN coating increases with increasing temperature and exceeds that of TiAlSiN/AlCrN. In contrast, the wear rate of TiAlSiN/AlCrN and AlCrN does not increase but decreases. In addition, regardless of whether the temperature is 500 °C or 800 °C, the wear rate of AlCrN coating is the highest compared to the other two.

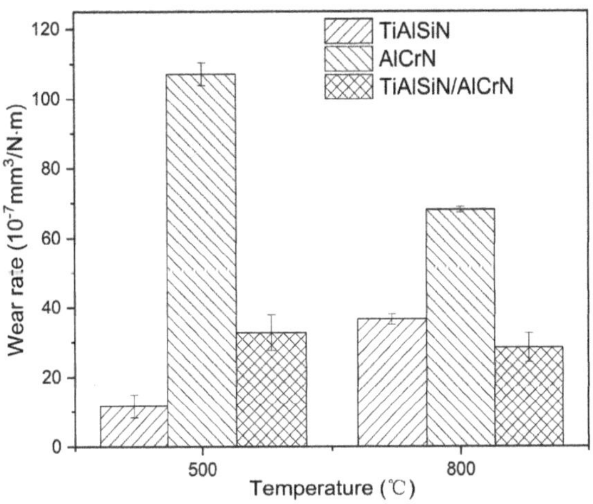

Figure 14. Wear rate of three coatings at high temperatures.

The explanation for the above results is firstly analyzed from the point of view of nanohardness and elastic modulus (Figure 4). The higher the H^3/E^{*2} value, the stronger the resistance to plastic deformation and wear [22,23]. The order of the magnitude of the H^3/E^{*2} values for each coating is TiAlSiN > TiAlSiN/AlCrN > AlCrN, which is consistent with the order of the resistance to wear of the coatings obtained from the friction tests at 500 °C. At 800 °C, although the temperature becomes higher, as AlCrN and TiAlSiN/AlCrN have higher onset oxidation temperatures than TiAlSiN, as shown in Figure 6, the H^3/E^{*2} value still dominates the wear resistance of these two coatings, and then AlCrN still maintains the highest wear rate at 800 °C. However, for TiAlSiN, although the ambient temperature of 800 °C does not reach its onset of oxidation temperature of 900 °C, in the actual friction tests, the temperature of the friction surface should be higher than the ambient temperature, meaning that the TiAlSiN coating has been oxidized and is thus unstable. The result is that the wear rate of TiAlSiN overtakes that of TiAlSiN/AlCrN coating.

4. Conclusions

(1) TiAlSiN/AlCrN multilayer coating shows a mixed crystal structure of TiAlSiN and AlCrN in the phase; the nanohardness, elastic modulus, and adhesion strength of the multilayer coating conform to the "law of mixtures," i.e., the above properties of TiAlSiN/AlCrN multilayer coating fall between those of the two monolayers. Benefiting from stress relief due to interfacial effects, TiAlSiN/AlCrN multilayer coating exhibits significantly lower residual stress than the two monolayer coatings.

(2) The order of oxidation resistance of the three coatings is TiAlSiN < TiAlSiN/AlCrN < AlCrN. The oxidation resistance of the TiAlSiN coating is attributed to the dense oxidation

product Al_2O_3 and the nanocomposite structure formed by Si. However, the loose porous oxide of Ti, TiO_2, negatively affects its oxidation resistance. The oxidation resistance of AlCrN coating benefits from the combined protection of the dense oxidation products Al_2O_3 and Cr_2O_3. The oxidation resistance of TiAlSiN/AlCrN multilayer coating inherits the advantages and disadvantages of the two monolayer coatings.

(3) The effect of the coatings' mechanical properties and oxidation resistance on the wear resistance depends on the temperature. At 500 °C, the order of wear resistance of the three coatings is consistent with the order of H^3/E^{*2} values, i.e., TiAlSiN > TiAlSiN/AlCrN > AlCrN; at 800 °C, the coating's oxidation resistance plays a crucial role, so the order of wear resistance changes into TiAlSiN/AlCrN > TiAlSiN > AlCrN due to TiAlSiN oxidizing before TiAlSiN/AlCrN.

Author Contributions: Experiments, J.L., Y.W. and J.H.; methodology, H.M. and X.D.; resources, Y.W.; data curation, J.H.; writing—original draft preparation, J.L.; writing—review and editing, H.M. and X.D.; visualization, J.W.; supervision and project administration, G.Y.; funding acquisition, X.D. and J.W. All authors have read and agreed to the published version of the manuscript.

Funding: This research was funded by the National Natural Science Foundation of China (52405189), the University-level Project of Guangzhou Panyu Polytechnic (2023KJ20 and 2021KJ12), the Guangzhou Basic and Applied Basic Research Project (202201011848), and the General Program of Philosophy and Social Science Planning of Guangdong Province (GD24CYS23).

Data Availability Statement: The original contributions presented in this study are included in the article. Further inquiries can be directed to the corresponding author.

Conflicts of Interest: The authors declare no conflicts of interest.

References

1. He, Q.; Saciotto, V.; DePaiva, J.M.; Guimaraes, M.C.; Kohlscheen, J.; Martins, M.M.; Veldhuis, S.C. Enhancing Tool Performance in High-Speed End Milling of Ti-6Al-4V Alloy: The Role of AlCrN PVD Coatings and Resistance to Chipping Wear. *J. Manuf. Mater. Process.* **2024**, *8*, 68. [CrossRef]
2. Zou, S.; Luo, Z.; Li, Y.; Yuan, L.; Tang, Y.; Zhou, J.; Li, H. Effect of Arc Current on the Microstructure of AlTiN-Coated Tools and Milling of 304 Stainless Steel. *Coatings* **2024**, *14*, 704. [CrossRef]
3. Zhao, F.; Wang, L.; Wang, X. Microstructure and properties of TiAlSiN ultra-hard coatings prepared by plasma immersion ion implantation and deposition with TiAlSi alloy cathodes. *Vacuum* **2020**, *174*, 109194. [CrossRef]
4. Das, C.R.; Rangwala, M.; Ghosh, A. Effect of Si contents on microstructure and mechanical characteristics of TiAlSiN thin film deposited by HiPIMS using different TiSi target compositions. *Surf. Coat. Technol.* **2024**, *476*, 130212. [CrossRef]
5. Das, C.R.; Rangwala, M.; Ghosh, A. Influence of substrate bias voltage on microstructure and mechanical characteristics of TiAlSiN coating deposited by High Power Impulse Magnetron Sputtering (HiPIMS). *Surf. Coat. Technol.* **2023**, *458*, 129351. [CrossRef]
6. Ji, L.; Liu, H.; Huang, C.; Li, S.; Yin, M.; Liu, Z.; Zhao, J.; Xu, L. Effect of TiAlSiN gradient structure design on mechanical properties and microstructure of coatings. *Surf. Coat. Technol.* **2025**, *496*, 131617. [CrossRef]
7. Yin, Z.; Wu, S.; Zhang, Y.; Yan, W.; Dai, S.; Peng, X.; Liao, B.; Zhang, X.; Wang, J.; Ouyang, X. A comparative study on the structure and properties of TiAlSiN coatings deposited by FCVA and HiPIMS. *J. Alloys Compd.* **2024**, *1005*, 175844. [CrossRef]
8. Wan, W.; Zhou, Q.; Liang, M.; Wang, P.; Rao, C.; Ji, S.; Fan, K.; Song, J. Study on the high temperature wear behavior of TiAlSiN coatings deposited on WC-TaC-Co cemented carbides. *Tribol. Int.* **2024**, *200*, 110115. [CrossRef]
9. Liu, Z.R.; Pei, F.; Chen, L.; Mayrhofer, P.H. Effect of Si-addition on structure and thermal stability of Ti-Al-N coatings. *J. Alloys Compd.* **2022**, *917*, 165483. [CrossRef]
10. Sousa, V.F.; Silva, F.; Alexandre, R.; Fecheira, J.; Silva, F. Study of the wear behaviour of TiAlSiN and TiAlN PVD coated tools on milling operations of pre-hardened tool steel. *Wear* **2021**, *476*, 203695. [CrossRef]
11. Liu, J.; Zhu, S.-S.; Deng, X.; Liu, J.-Y.; Wang, Z.-P.; Qu, Z. Cutting performance and wear behavior of AlTiN-and TiAlSiN-coated carbide tools during dry milling of Ti–6Al–4V. *Acta Metall. Sin. (Engl. Lett.)* **2020**, *33*, 459–470. [CrossRef]
12. Kameneva, A.; Kichigin, V. Corrosion, wear, and friction behavior of a number of multilayer two-, three-and multicomponent nitride coatings on different substrates, depending on the phase and elemental composition gradient. *Appl. Surf. Sci.* **2019**, *489*, 165–174.

13. Vorontsov, A.; Filippov, A.; Shamarin, N.; Moskvichev, E.; Novitskaya, O.; Knyazhev, E.; Denisova, Y.; Leonov, A.; Denisov, V.; Tarasov, S. High-Temperature Oxidation of CrN/ZrN Multilayer Coatings. *Metals* **2022**, *12*, 1746. [CrossRef]
14. Liu, H.; Yang, F.-C.; Tsai, Y.-J.; Wang, X.; Li, W.; Chang, C.-L. Effect of modulation structure on the microstructural and mechanical properties of TiAlSiN/CrN thin films prepared by high power impulse magnetron sputtering. *Surf. Coat. Technol.* **2019**, *358*, 577–585.
15. Chang, Y.-Y.; Chang, C.-E. Mechanical properties and tribological performance of multilayered AlCrBN/AlTiBN coatings. *Surf. Coat. Technol.* **2025**, *496*, 131691.
16. Xiao, B.; Zhang, T.F.; Guo, Z.; Li, Z.; Fan, B.; Chen, G.; Xiong, Z.; Wang, Q. Mechanical, oxidation, and cutting properties of AlCrN/AlTiSiN nano-multilayer coatings. *Surf. Coat. Technol.* **2022**, *433*, 128094.
17. Liu, J.; Wang, Y.; Liu, G.; Hua, J.; Deng, X. Properties and Performance of TiAlSiN and AlCrN Monolayer and Multilayer Coatings for Turning Ti-6Al-4V. *Coatings* **2023**, *13*, 1229. [CrossRef]
18. Chen, P.-Y.; Wang, W.-C.; Wu, Y.-T. Experimental investigation of thin film stress by Stoney's formula. *Measurement* **2019**, *143*, 39–50.
19. Yousaf, M.; Pelenovich, V.; Yang, B.; Liu, C.; Fu, D. Effect of bilayer period on structural and mechanical properties of nanocomposite TiAlN/MoN multilayer films synthesized by cathodic arc ion-plating. *Surf. Coat. Technol.* **2015**, *282*, 94–102.
20. Mei, H.; Yan, K.; Wang, R.; Peng, W.; Huang, K.; Shi, J.; Zhang, D.; Gong, W.; Ren, F.; Wang, Q. Microstructure and mechanical properties of nanomultilayered AlTiN/Cu coatings prepared by a hybrid system of AIP and PDCMS. *Ceram. Int.* **2023**, *49*, 226–235.
21. Kimura, A.; Kawate, M.; Hasegawa, H.; Suzuki, T. Anisotropic lattice expansion and shrinkage of hexagonal TiAlN and CrAlN films. *Surf. Coat. Technol.* **2003**, *169*, 367–370. [CrossRef]
22. Mayrhofer, P.H.; Mitterer, C.; Musil, J. Structure–property relationships in single- and dual-phase nanocrystalline hard coatings. *Surf. Coat. Technol.* **2003**, *174–175*, 725–731. [CrossRef]
23. Philippon, D.; Godinho, V.; Nagy, P.M.; Delplancke-Ogletree, M.P.; Fernández, A. Endurance of TiAlSiN coatings: Effect of Si and bias on wear and adhesion. *Wear* **2011**, *270*, 541–549. [CrossRef]
24. Kumar, P.N.; Kannan, S. Sequential elucidation of the β-$Ca_3(PO_4)_2$/TiO_2 composite development from the solution precursors. *Dalton Trans.* **2017**, *46*, 3229–3239.
25. Shi, T.; Duan, Y.; Lv, K.; Hu, Z.; Li, Q.; Li, M.; Li, X. Photocatalytic oxidation of acetone over high thermally stable TiO_2 nanosheets with exposed (001) facets. *Front. Chem.* **2018**, *6*, 175.
26. Xin, L.; Chen, Q.; Teng, Y.; Wang, W.; Sun, A.; Zhu, S.; Wang, F. Effects of silicon and multilayer structure of TiAl (Si) N coatings on the oxidation resistance of Ti6Al4V. *Surf. Coat. Technol.* **2013**, *228*, 48–58.
27. Chang, Y.-Y.; Hsiao, C.-Y. High temperature oxidation resistance of multicomponent Cr–Ti–Al–Si–N coatings. *Surf. Coat. Technol.* **2009**, *204*, 992–996. [CrossRef]
28. Pfeiler, M.; Zechner, J.; Penoy, M.; Michotte, C.; Mitterer, C.; Kathrein, M. Improved oxidation resistance of TiAlN coatings by doping with Si or B. *Surf. Coat. Technol.* **2009**, *203*, 3104–3110.
29. Vennemann, A.; Stock, H.R.; Kohlscheen, J.; Rambadt, S.; Erkens, G. Oxidation resistance of titanium–aluminium–silicon nitride coatings. *Surf. Coat. Technol.* **2003**, *174–175*, 408–415.
30. Zhang, X.D.; Chen, L.; Wang, G.Y.; Hu, C.; Du, J.W. Influence of CrAlN layers on the microstructure, thermal stability, oxidation and corrosion resistance of AlN/CrAlN multilayers. *Mater. Charact.* **2024**, *214*, 114051.
31. Reiter, A.; Mitterer, C.; Sartory, B. Oxidation of arc-evaporated $Al_{1-x}Cr_xN$ coatings. *J. Vac. Sci. Technol. A* **2007**, *25*, 711–720. [CrossRef]
32. Xu, Y.; Chen, L.; Liu, Z.; Pei, F.; Du, Y. Influence of Ti on the mechanical properties, thermal stability and oxidation resistance of Al–Cr–N coatings. *Vacuum* **2015**, *120*, 127–131. [CrossRef]
33. Jie, S.; An, S.; Pengjiao, Z.; Junzhou, L. Microstructure, mechanical performance, thermal stability, and oxidation resistance of AlCrN/AlCrBN nano-multilayer coating. *Surf. Coat. Technol.* **2024**, *493*, 131301.
34. Polcar, T.; Cavaleiro, A. High-temperature tribological properties of CrAlN, CrAlSiN and AlCrSiN coatings. *Surf. Coat. Technol.* **2011**, *206*, 1244–1251. [CrossRef]

nanomaterials

MDPI

Article

Investigating the Effects of Long-Term Ambient Air Storage on the Sliding Properties of N-Alloyed $MoSe_2$ Coatings

Talha Bin Yaqub [1,2,*,†], Irfan Nadeem [1,†], Muhammad Aneeq Haq [3], Muhammad Yasir [4], Albano Cavaleiro [2] and Mitjan Kalin [1]

1 Laboratory for Tribology and Interface Nanotechnology (TINT), Faculty of Mechanical Engineering, University of Ljubljana, Askerceva 6, 1000 Ljubljana, Slovenia
2 Department of Mechanical Engineering, CEMMPRE, ARISE, University of Coimbra, Rua Luis Reis Santos, 3030-788 Coimbra, Portugal; albano.cavaleiro@dem.uc.pt
3 School of Chemical and Materials Engineering (SCME), National University of Sciences and Technology (NUST), H-12, Islamabad 46000, Pakistan; aneeq.haq@scme.nust.edu.pk
4 Department of Fire Safety Engineering and Management ICEM, Muscat 2511, Oman
* Correspondence: talhabin.yaqub@fs.uni-lj.si
† These authors contributed equally to this work.

Abstract: Transition metal dichalcogenide coatings have emerged as potential candidates for terrestrial and aerospace mobility applications. Among these, the alloyed $MoSe_2$ coatings have displayed promising results while sliding in diverse environments. N-alloyed $Mose_2$ coatings provide the additional benefit of overcoming the impact of PVD compositional variations on dry sliding, making them promising solid lubricants for mobility-sector applications. However, the impact of long-term storage has never been investigated for this rarely studied solid-lubricant system. This study investigates the tribological performance of direct current magnetron sputtered MoSeN coatings after 40 months of storage in an ambient atmosphere. Sliding tests were conducted under conditions consistent with pre-storage conditions. The results showed that coatings with 0 at. %, 22 at. %, 33 at. %, and 35 at. % N-alloying exhibited COF values nearly identical to the pre-storage results, with only a negligible increase in ~0.01. Similarly, all coatings displayed specific wear rates in the range of 10^{-7}, aligning with earlier findings. The obtained results show that the sliding performance of MoSeN coatings does not deteriorate over time, highlighting their suitability for critical aerospace applications, where components and assembled parts may be stored for years before launching into space or in actual applications.

Keywords: MoSeN coatings; storage life cycle; solid lubricant; friction and wear; aerospace industry

1. Introduction

In past few decades, considerable efforts have been made to develop a solid lubricant-based coating system that can mitigate the adverse effects being faced due to liquid lubrication in both terrestrial and non-terrestrial atmospheres [1–3]. Thus, the modern research is aggressively focused on the optimization of tribo-mechanisms for reduction in friction and wear in mobility sectors via a sustainable solid lubricant coating system. This research has brought forth promising results, and a few of these coatings are already in use with continuing efforts for their efficiency enhancement [4–6].

Moreover, similar efforts resulted in emergence of a potentially suitable systems based on transition metal dichalcogenide (TMD) coatings [7,8]. Basically, TMDs have a layered structure which helps to provide easy shear properties and thus fulfill the requirements of low friction materials [9,10]. TMDs have been utilized since 1960s [11]. Significant efforts have been made to enhance the properties of TMD based coatings for efficient sliding. These include preparation by ion-beam-assisted depositions, RF, DC, Diode, and magnetron sputtering sources [12–17]. Similarly, refs. [18–22] report the efforts carried out to prepare TMD coatings with enhanced characteristics after alloying of metals and their oxides or preparing them in the form of multi-layers. Nevertheless, the reports published in ref. [8,14,23] provide thorough understanding of various approaches that were used to synthesize optimized TMD coatings until the late 19th century, as well as the critical performance results in various sliding conditions. However, despite promising results, developing a suitable solid lubricant that can provide long-term lubrication was an issue. The sputtered coatings based on pure TMDs are actually not capable of resisting wear due to their low load-bearing capabilities [24,25]. Additionally, the porous morphology is not capable of resisting moisture and similar environmental attacks, resulting in adverse sliding performance and increments of COF [22,26,27]. In an effort to combine different types of 2D materials for enhanced sliding efficiency and to overcome the adverse effects of pure TMDs, Voevodin combined the pure WS_2 coatings with WC and DLCs to enhance the environmental resistance and load-bearing capabilities of these coatings [28]. Productive results were obtained, and the coating sliding performance improved substantially. This was followed by thorough investigations of these coatings where efforts were made that involved alloying with different metals [29–33] and non-metals [34–37] explored to enhance the compactness, reduce porosity, and improve load-bearing capabilities for effective enhancement in wear resistance. The efforts resulted in fruitful results, with the best performance being achieved with both C and N alloying of different TMDs (e.g., MoS_2, WS_2, and $MoSe_2$).

Among these coating systems, the direct current (DC) magnetron sputtered N-alloyed $MoSe_2$ (i.e., MoSeN) system has been recently developed and reported by current authors [38]. These MoSeN coatings displayed promising sliding properties in ambient air and proved suitable for 3D complex parts as they efficiently overcame the impacts of compositional variations on the sliding efficiency typically faced in PVD sputtered coatings.

Despite the development of different coating systems that can provide low friction in different environments, the scientific community pays less attention to a crucial aspect, i.e., the storage life analysis of these under-development coatings. In the urge to develop a long-lasting and durable solution, it is critically important to see the performance of solid lubricant coatings over time. This is vital as the coatings may perform above expectations soon after synthesis; however, long-term storage may degrade their efficient sliding performances. There are various means of degradation of the properties like (i) the storage conditions and atmosphere, (ii) the moisture present in the ambient atmosphere, and (iii) the temperature fluctuations that are experienced in seasonal variations, etc. All these factors can contribute to the changes in surface properties like impurities introduction, built-up of atomic layer oxides, or modifications of the structure of the coatings due to corrosion or environmental degradation (e.g., due to moisture attack), etc. It is very common in industries that coated parts often remain in inventory for years before being employed in machines. Thus, understanding consistent efficiency, regardless of the time elapsed before use, is crucial. As per the best knowledge of the present authors, only a few studies report the analysis of TMD coatings after some years of storage. For example, a study by Gustavsson et al. [39] mentioned only the analysis of chemical composition and structure of Mo-Se-C coatings after a gap of 5 years, and no significant variation was achieved in these properties. Similarly, the aging effects of TMDs and the growth of thin oxide layers have been reported in various studies referred to in refs. [40–42].

To the best of our knowledge, the impact of long-term storage on sliding properties of alloyed-TMD coatings, especially Mo-Se-N, has never been explored. Thus, considering the need to assess the sliding properties after long-term storage, MoSeN coatings deposited by current authors in April 2021 were re-evaluated after 40 months of storage and compared with previously measured results. Over a period of 40 months, these coatings were stored under ambient air conditions in plastic bags without any additional precautions. Therefore, the aim of this study is to emphasize the critical need to assess the storage durability of the coatings, particularly for demanding applications in industries such as automotive and aerospace, where consistent performance over an extended (storage) period is crucial.

2. Experimental Procedure

The coatings were synthesized in a semi-industrial scale deposition chamber by a N_2 gas reactive magnetron sputtering of a DC-powered $MoSe_2$ target in an Ar atmosphere. The depositions were carried out at a fixed chamber pressure of 0.27 Pa. The N_2 gas flow was varied from 0 to 5 sccm to allow the deposition of a series of coatings with varying nitrogen content. Besides the $MoSe_2$ target, a Cr target was also utilized for the deposition of the Cr interlayer and Cr-MoSeN gradient layer. The total deposition time for interlayer and gradient layers was 15 min each while the final coating deposition time was varied to achieve coatings of approx. 2 μm thickness. Mirror-polished DIN 100Cr6 (AISI-52100) steel (Ø 25 × 7 mm) samples having roughness (Ra) less than 0.02 μm and ~9.8 GPa hardness were used as substrates. The substrates were ultrasonically cleaned using ethanol for 15 min before deposition. A set of four coatings, alloyed with 0–35 at. % N, were deposited and stored for 40 months. The results of tests performed soon after coating deposition (i.e., before storage) are reported in ref. [38].

Post-storage sliding tests were carried out using a reciprocating sliding tribometer (UMT-2, Bruker, Billerica, MA, USA) for 1200 s under ambient conditions (RH = ~30–35%) against a 6 mm 100Cr6 ball. The stroke length and sliding frequency were fixed to 5 mm and 10 Hz, respectively. The tests were performed at an average sliding speed of 0.1 m/s during the stroke and an applied normal load of 3.5 N, corresponding to an initial Hertzian contact pressure of 1 GPa. These parameters were chosen to replicate the sliding conditions used in the previous study conducted 40 months earlier [38]. The tests were repeated three times, and the COF shown was systematically calculated based on the evolutions. This means that initially, the results from three repetitions were obtained. Then, the steady state regions of the COF evolution of each repetition were selected to calculate the average COF and standard deviation of that repetition. Finally, the average steady state COF and respective standard deviations of all repetitions were used to calculate the final average COF and deviations (errors). Thus, the bar charts represent these final average values, and the comparison has been made among the achieved final COF values shown in the bar chart.

After testing, ball and disk worn regions were examined using a digital optical microscope (Hirox-HRX-01, Kyoto, Japan). For disks, the wear profiles at three different zones of wear track from each repetition were measured using a white-light interferometer (Bruker Contour GT-K0, Billerica, MA, USA). These wear profiles were used to calculate the worn area. The worn area and stroke length were then used to calculate the total wear volume. Finally, the wear volume, applied load, and sliding distances were used to calculate specific wear rates in mm^3/Nm units. The average values were then calculated and reported in this work. Likewise, specific wear rates of the steel counterparts were determined by measuring the worn scar diameters observed under the optical microscope.

The as-deposited coatings and disk wear tracks were also analyzed in Raman spectroscopy using a confocal Raman microscope (Horiba Xplora Plus, Kyoto, Japan) for the investigation of the sliding-induced structural changes and the possible mechanism behind the frictional results obtained in this work. The Raman analysis was performed using a 512 nm laser and the acquisition conditions of 1200 grating, 6 repetitions of 15 secs each, and 90% filter were chosen to avoid damage to the coatings and wear tracks.

3. Results and Discussion

3.1. Fundamental Properties of the Coatings

The fundamental characteristics of the coatings are detailed in ref. [38]. Briefly, the coatings contained N concentrations ranging from 0 to 35 at.%. This resulted in Se/Mo ratio ranging from 1.6 to 2.0, which decreased with increasing N content (Figure 1). The coatings are denominated as mentioned in Figure 1, i.e., N0, N1, N2, and N3 correspond to 0 at. % N, 21.8 at. % N, 33.1 at. % N, and 35 at. % N, respectively. The morphology of coatings transitioned from columnar and porous for pure coating to dense and compact with N-alloying. In terms of crystal structure, the pure coating (N0) exhibited crystalline peaks corresponding to the (002) basal planes as well as (100) and (10L) peaks. Whereas, N incorporation enhanced the amorphousness, and the coatings became nano-crystalline. The pure coating exhibited a hardness of 1.1 GPa, which increased to about 5 GPa with N addition (Figure 1). Overall, except for the pure coating, all other MoSeN coatings exhibited almost similar morphology, crystal structure, and hardness values, regardless of N content and Se/Mo ratios. These consistent properties, despite compositional variations, are advantageous for the industrial application of PVD sputtered coatings. This is because PVD sputtering is a highly line of sight deposition method [43], and, thus, the compositional variations for depositions on 3D parts is thus very common.

Coating	N content (at. %)	Se/Mo	Hardness (GPa)	Crystal Structure
N0	-	2.0	1.1±0.1	Crystalline with (002) & (10L) peaks
N1	21.8±0.5	1.8	4.6±0.2	Nano-composite with broad (10L) peaks
N2	33.1±0.4	1.6	4.7±0.3	
N3	35.0±0.6	1.6	5.1±0.3	

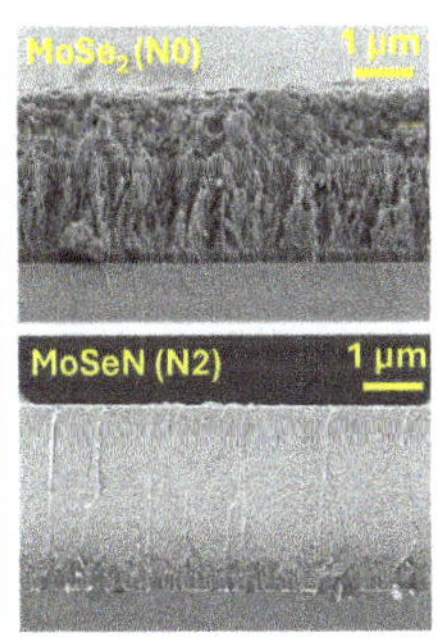

Figure 1. Overview of fundamental characteristics of MoSeN coatings—composition, crystal structure, and SEM morphology micrographs (N2 micrograph is shown as representative of all MoSeN coatings).

3.2. Tribological Results

The average COF results after 40 months of storage are shown in Figure 2a. For all coatings, initially, the COF decreased and then increased in the running-in phase to finally reach the steady-state sliding zone. The pure N0 coating exhibited an average steady-state COF value of 0.081. With N-alloying, the N1 coating showed a steady state average COF of 0.062. Similarly, the N2 coating's average steady-state COF was 0.061. For N3 coating, the COF never reached a stable steady state zone, and it continued to vary. For example, in one of the repetitions, the minimum COF was around 0.055 (at 130 s), and then it continued to increase till the end of the test. Thus, for this coating, the average was calculated from the point of minimum COF till the end of the test. So, the average COF calculated was 0.076. The COF evolution comparison in pre- and post-storage tests is shown in Figure 2c and Figure 2d, respectively.

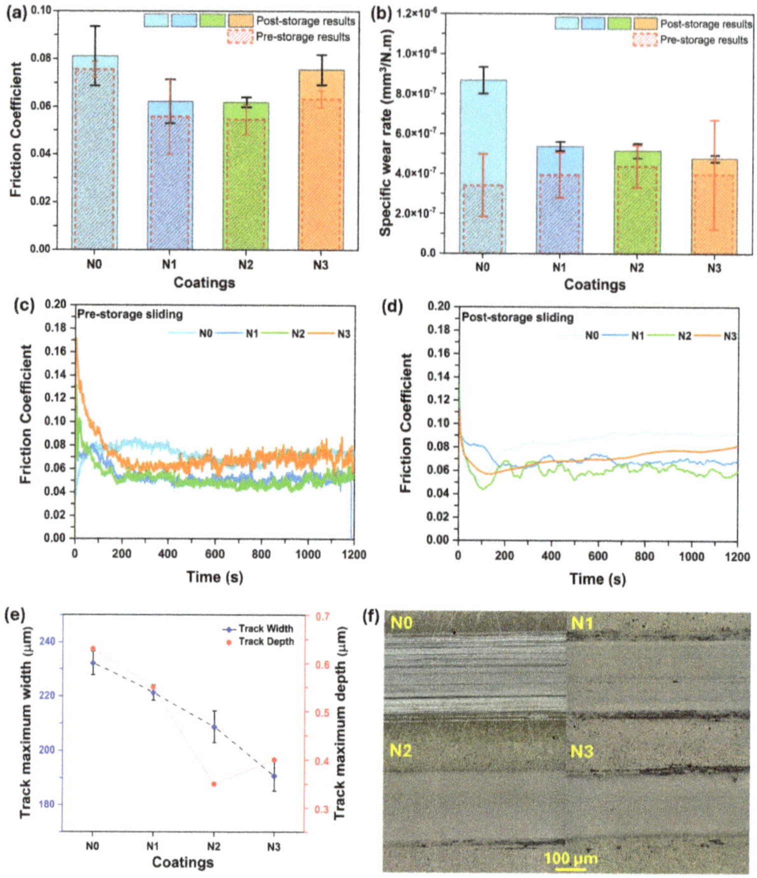

Figure 2. Tribological performance of MoSeN coatings, (**a**) post-storage average friction coefficients, (**b**) post-storage average specific wear rates, (**c**) pre-storage friction coefficients evolution, (**d**) post-storage friction coefficients evolution, (**e**) maximum wear-tracks width and depth from post-storage results, and (**f**) optical images of wear tracks from post-storage results.

Thus, from the obtained results, it was observed that for N-alloyed coatings, the N3 coating displayed slightly unstable/fluctuating COF and slightly increased value. Overall, all coatings exhibited a minor increase in the COF in comparison to the results obtained immediately after deposition (40 months ago). It should be noted that this minor increase can be attributed to the build-up of a small oxidation layer on the very small vicinity of the coating surface during this long-storage duration. It is very common that the atmospheric oxygen may interact with the surface of materials during their storage in a normal atmosphere. However, this interaction has not shown a significant impact on the COF of the coatings even after 40 months, which seems impressive.

The specific wear rates of the coated disks are shown in Figure 2b. The N0 coating exhibited the highest specific wear rate of 8.67×10^{-7} mm^3/Nm. The wear rate decreased to 5.36×10^{-7} mm^3/Nm for N1 coating. With further increments in the N content, the specific wear rates further decreased to 5.13×10^{-7} mm^3/Nm for N2 coating and 4.76×10^{-7} mm^3/Nm for N3 coating. The higher specific wear rate of the N0 coatings compared to N-alloyed coatings is due to its porous morphology and the low load-bearing capability of pure sputtered TMD coatings. The porous morphology makes the pure TMD coatings more prone to degradation because of environmental attacks, e.g., moisture [44,45]. Basically, in such cases, the O or moisture from the atmosphere can passivate the vacant dangling bonds and result in increased friction and wear of the coatings. For the N-alloyed coatings, wear rates quite marginally decreased with increasing N. However, the decrease as compared to the pure coating is more. This decrease is attributed to compact morphology and more environmental attack resistance.

Similarly, the slightly higher specific wear rates of the coatings as compared to the pre-storage conditions may be attributed to the slight oxidation of the outer surface which can cause the formation of oxides that can create abrasiveness. Moreover, the more profound aspect is the fact that a different specific wear rate measurement mechanism, i.e., a 3D white light optical interferometer (in VXI mode), with a greater resolution was used in the current study, whereas a low-resolution stylus profilometry was utilized in previous work. In pre-storage measurements, stylus profilometry was employed, which has less sensitivity as compared to the white light interferometer used in the current study. Nonetheless, the wear rate still lies in the 10^{-7} range, which is a promising result.

Figure 2e displays that the maximum wear track depths observed for N0, N1, N2, and N3 coatings were 0.63, 0.58, 0.34, and 0.42 μm, respectively. The optical images of coating wear tracks are shown in Figure 2f. All tracks showed abrasive wear marks and displayed features distinct from the as-deposited coatings. As per the previous experience and reported results, the tracks were covered with adhered materials, most likely $MoSe_2$ tribolayers. The degree of coverage varied across coatings, with coverage decreasing with increasing nitrogen content. This decrease is directly related to the Se/Mo ratio, with higher N content reducing the availability of $MoSe_2$ for tribolayer formation and coverage.

Similarly, the 3D surface topography of wear tracks along-with 2D profiles are presented in Figure 3. Agreeing with the specific wear rate results, the pure reference coating displayed the highest wear and this decreased with the introduction of N in the coatings. It is clear that the sliding was always taking place in the outer coating, and it never reached the interlayer, even for the highest wear track depth zone. Moreover, the 3D topography and 2D profilometer results are in accordance with the optical images displayed in Figure 2f.

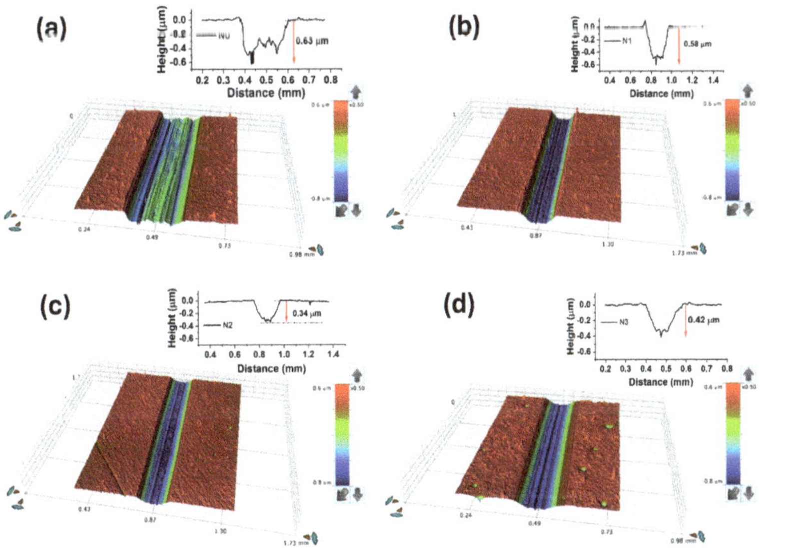

Figure 3. Three-dimensional surface topography and 2D profiles of wear tracks, (**a**) N0 coating, (**b**) N1 coating, (**c**) N2 coating and, (**d**) N3 coating.

The specific wear rates and wear scars of the steel counterparts are shown in Figure 4a,b. The N0 coating steel counterpart showed no wear and was covered by a thick layer of adhered material, which correlates with the high disk wear rate. Basically, the N0 coating was softer and, due to the high wear of the soft coating, enough material was transferred to protect the counterpart. Specific wear rate for the N1 coating counterpart was 2.48×10^{-8} mm^3/Nm, which minorly increased to 2.72×10^{-8} mm^3/Nm for the N2 coating counterpart while it then decreased to 2.14×10^{-8} mm^3/Nm for the N3 coating counterpart.

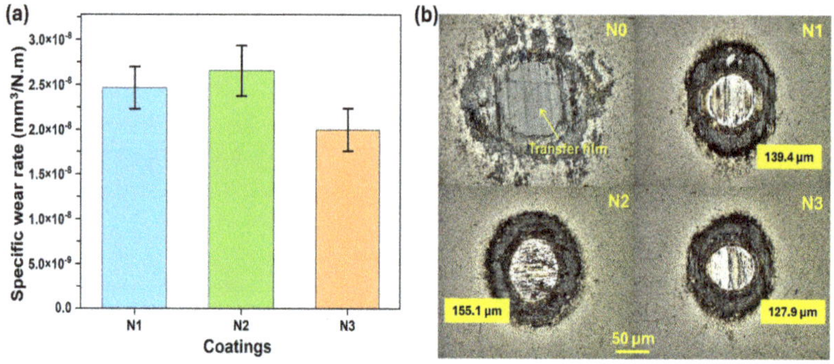

Figure 4. Sliding results—(**a**) specific wear rates of counter bodies (balls) and (**b**) optical images of ball wear scars.

3.3. Raman Spectroscopy Analysis of Wear Tracks

The Raman spectroscopy analysis of the wear tracks was performed to investigate the reasons behind the low friction of the coatings as well as the role of TMD phase during sliding. The analysis was performed on the as-deposited surfaces and in the middle of the wear tracks. The results for the as-deposited coatings and wear tracks are shown in Figure 5a and Figure 5b, respectively. From the as-deposited coatings zone results, it is clear that the pure N0 coating displayed well defined crystalline peaks of $MoSe_2$ while the N-alloyed as-deposited coatings did not exhibit well-defined peaks. These observations are consistent with the crystal structure results (see Figure 1). However, the analysis performed inside the wear tracks revealed distinct crystalline $MoSe_2$ peaks in all coatings. This increase in crystallinity after sliding indicates tribo-induced crystallization as well as chemical changes in the contact zone. This crystallization results in the formation of an easy shear tribolayer of $MoSe_2$, driven by reorientation and transfer layer mechanisms, as observed in MoSeC coatings [46]. Thus, in agreement with refs. [46,47], the formation of $MoSe_2$ tribolayers is the key to low friction in the present case. These Raman spectroscopy results thus confirm the hypothesis mentioned above in the optical image results that the tracks features are different than the as-deposited coatings and the tracks are covered with an adhered material (transfer film).

In summary, it is clear that the 40 months ambient air storage had no significant impact on the performance or shelf-life of the coatings. All coatings except N0 exhibited similar friction and wear behavior with very minor increase in comparison to the results obtained immediately after deposition (40 months ago). These coatings remained in an ambient atmosphere with seasonal variations in humidity and minor temperature fluctuations in winters and summers. Thus, the results depict a minor difference in ~0.01 in COF and within 10^{-7} mm^3/Nm range wear rates. Moreover, the slight differences in specific wear rates can be attributed to the use of a white-light interferometer in this study, offering higher resolution than the stylus profilometer used in the pre-storage tests. Therefore, these coatings are quite stable and suitable for industrial applications requiring long-term storage. Furthermore, it is noteworthy that the consistent sliding performance of N-alloyed coatings, regardless of compositional variations, further reinforces the previous findings. The ability to maintain a similar COF across different compositions is a significant achievement, addressing the drawbacks of PVD sputtering and making MoSeN coatings particularly favorable for industries relying on PVD sputtering technology.

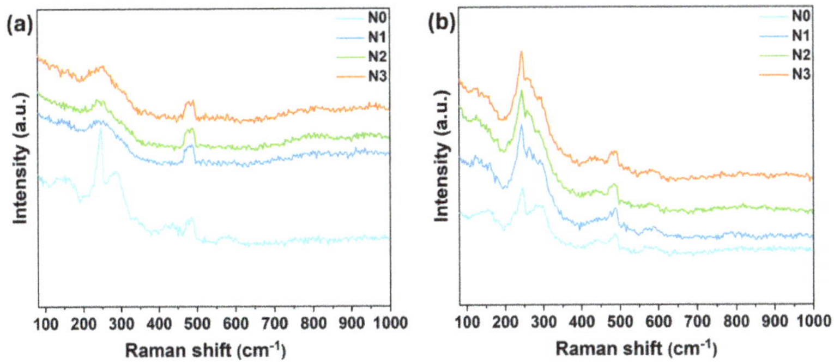

Figure 5. Raman spectra acquired on the coated disks. (**a**) As-deposited coating zone, and (**b**) inside the wear tracks ($MoSe_2$ peaks region = 150–350 cm^{-1}).

4. Conclusions

This research investigates the sliding characteristics of MoSeN coatings after a long-term storage of 40 months. The MoSeN coatings with N content ranging from 0 to 35 at. % were deposited using reactive DC magnetron sputtering. The sliding tests performed after 40 months of storage showed results very similar to pre-storage tests and only minor variations were observed. The pure N0 coating displayed a COF of 0.081 and a wear rate of 8.67×10^{-7} mm^3/Nm, while all MoSeN coatings displayed COF values between 0.061 and 0.076, with wear rates ranging from 4.76 to 5.35×10^{-7} mm^3/Nm. Two-dimensional profiles and 3D topography results demonstrated that the coating did not delaminate, and the sliding occurred within the outer coating. Similarly, Raman spectroscopy analysis of virgin coating and wear tracks concluded that crystallized $MoSe_2$ signals were detected only inside the wear tracks and not from virgin MoSeN coating surfaces. This proved that the low friction characteristics are governed by $MoSe_2$ tribolayers. These findings demonstrate that globally, the coatings' performance is unaffected by long-term storage, making them suitable for industries requiring extended storage periods before implementation.

Author Contributions: Conceptualization, T.B.Y.; Methodology, T.B.Y., I.N. and M.A.H.; Validation, T.B.Y.; Formal analysis, T.B.Y., I.N. and M.A.H.; Investigation, T.B.Y. and M.Y.; Resources, A.C. and M.K.; Data curation, T.B.Y. and I.N.; Writing—original draft, T.B.Y. and I.N.; Writing—review & editing, A.C. and M.K.; Visualization, T.B.Y.; Supervision, A.C. and M.K.; Project administration, A.C. and M.K.; Funding acquisition, T.B.Y., A.C. and M.K. All authors have read and agreed to the published version of the manuscript.

Funding: This research is funded by the Slovenian Research Agency ARIS under the Research Core Funding Programme No. P2-0231 and Marie-Sklodowska Curie COFUND—Seal of Excellence No. 5100-237/2023-7 (5 February 2024). The work is also sponsored by national funds through FCT—Fundação para a Ciência e a Tecnologia, under projects UID/00285—Centre for Mechanical Engineering, Materials and Processes and LA/P/0112/2020.

Data Availability Statement: Data is contained within the article.

Acknowledgments: The authors acknowledge the support of Laboratory for Tribology and Interface Nanotechnology (TINT), Faculty of Mechanical Engineering, University of Ljubljana and Centre for Mechanical Engineering, Materials and Processes, University of Coimbra.

Conflicts of Interest: The authors declare no competing financial or personal interests.

References

1. Fitch, E.C. Fluid Chemical Stability. In *Proactive Maintenance for Mechanical Systems*; Elsevier: London, UK, 1992; pp. 99–126.
2. Ciulli, E. *Tribology and Sustainable Development Goals*; Quaglia, G., Gasparetto, A., Petuya, V., Carbone, G., Eds.; Springer International Publishing: Cham, Switzerland, 2022; pp. 438–447.
3. Technical Committee of Petroleum Additive Manufacturers in Europe. *Lubricant Additives and the Environment*; Atc Document 49; Technical Committee of Petroleum Additive Manufacturers in Europe: Brussels, Belgium, 2007.
4. Donnet, C.; Erdemir, A. *Diamond-like Carbon Films: A Historical Overview*; Springer: Berlin/Heidelberg, Germany, 2008; ISBN 9780387302645.
5. Hilton, M.R.; Fleischauer, P.D. Applications of Solid Lubricant Films in Spacecraft. *Surf. Coat. Technol.* **1992**, *54–55*, 435–441. [CrossRef]
6. Aouadi, S.M.; Gao, H.; Martini, A.; Scharf, T.W.; Muratore, C. Lubricious Oxide Coatings for Extreme Temperature Applications: A Review. *Surf. Coat. Technol.* **2014**, *257*, 266–277. [CrossRef]
7. Lansdown, A.R. *Molybdenum Disulphide Lubrication*; Elsevier: Amsterdam, The Netherlands, 1999; ISBN 9780080536927.
8. Roberts, E.W. Thin Solid Lubricant Films in Space. *Tribol. Int.* **1990**, *23*, 95–104. [CrossRef]
9. Yaqub, T.B.; Vuchkov, T.; Evaristo, M.; Cavaleiro, A. DCMS Mo-Se-C Solid Lubricant Coatings—Synthesis, Structural, Mechanical and Tribological Property Investigation. *Surf. Coat. Technol.* **2019**, *378*, 124992. [CrossRef]
10. Yaqub, T.B.; Al-rjoub, A.; Cavaleiro, A.; Fernandes, F. Exploring the Industrial Implementation of W-S-N Coatings: A Detailed Study of the Synthesis, Compositional, Structural, Mechanical and Multi-Environment Lubrication Properties. *J. Mater. Res. Technol.* **2022**, *18*, 547–563. [CrossRef]
11. Voevodin, A.A.; Muratore, C.; Aouadi, S.M. Hard Coatings with High Temperature Adaptive Lubrication and Contact Thermal Management: Review. *Surf. Coat. Technol.* **2014**, *257*, 247–265. [CrossRef]
12. Bolster, R.N.; Singer, I.L.; Wegand, J.C.; Fayeulle, S.; Gossett, C.R. Preparation by Ion-Beam-Assisted Deposition, Analysis and Tribological Behavior of MoS2 Films. *Surf. Coat. Technol.* **1991**, *46*, 207–216. [CrossRef]
13. Sliney, H.E. Solid Lubricant Materials for High Temperatures—A Review. *Tribol. Int.* **1982**, *15*, 303–315. [CrossRef]
14. Roberts, E.W. Ultralow Friction Films of MoS2 for Space Applications. *Thin Solid Films* **1989**, *181*, 461–473. [CrossRef]
15. Stewart, T.B.; Fleischauer, P.D. Chemistry of Sputtered Molybdenum Disulfide Films. *Inorg. Chem.* **1982**, *21*, 2426–2431. [CrossRef]
16. Sekine, T.; Izumi, M.; Nakashizu, T.; Uchinokura, K.; Matsuura, E. Raman Scattering and Infrared Reflectance in 2H-MoSe2. *J. Phys. Soc. Jpn.* **1980**, *49*, 1069–1077. [CrossRef]
17. Bichsel, R.; Levy, F. Morphological and Compositional Properties of MoSe2 Films Prepared by R.F Magnetron Sputtering. *Thin Solid Films* **1984**, *116*, 367–372. [CrossRef]
18. Mikhailov, S.; Savan, A.; Pflüger, E.; Knoblauch, L.; Hauert, R.; Simmonds, M.; Swygenhoven, H. Van Morphology and Tribological Properties of Metal (Oxide)–MoS2 Nanostructured Multilayer Coatings. *Surf. Coat. Technol.* **1998**, *105*, 175–183. [CrossRef]
19. Wahl, K.J.; Dunn, D.N.; Singer, I.L. Wear Behavior of Pb–Mo–S Solid Lubricating Coatings. *Wear* **1999**, *230*, 175–183. [CrossRef]
20. Simmonds, M.C.; Savan, A.; Van Swygenhoven, H.; Pflüger, E.; Mikhailov, S. Structural, Morphological, Chemical and Tribological Investigations of Sputter Deposited MoSx/Metal Multilayer Coatings. *Surf. Coat. Technol.* **1998**, *108–109*, 340–344. [CrossRef]
21. Spalvins, T. Frictional and Morphological Properties of Au-MoS2 Films Sputtered from a Compact Target. *Thin Solid. Films* **1984**, *118*, 375–384. [CrossRef]
22. Hilton, M.R. Fracture in MoS2 Solid Lubricant Films. *Surf. Coat. Technol.* **1994**, *68–69*, 407–415. [CrossRef]
23. Fusaro, R.L. *Lubrication and Failure Mechanisms of Molybdenum Disulfide Films*; National Aeronautics and Space Administration, Scientific and Technical Information Office: Washington, DC, USA, 1978.
24. Polcar, T.; Cavaleiro, A. Self-Adaptive Low Friction Coatings Based on Transition Metal Dichalcogenides. *Thin Solid Films* **2011**, *519*, 4037–4044. [CrossRef]
25. Fu, Y.; Wang, Q.; Guo, L.; Zhao, X.; Jiang, D.; Gao, X.; Weng, L.; Sun, J.; Hu, M.; Wang, D. Interlayer Texturing for Improving the Tribological Property in Vacuum of Highly Crystallized Molybdenum Disulfide Film. *Vacuum* **2023**, *213*, 112100. [CrossRef]
26. Wang, D.-Y.; Chang, C.-L.; Ho, W.-Y. Microstructure Analysis of MoS2 Deposited on Diamond-like Carbon Films for Wear Improvement. *Surf. Coat. Technol.* **1999**, *111*, 123–127. [CrossRef]
27. Voevodin, A.A.; Zabinski, J.S. Nanocomposite and Nanostructured Tribological Materials for Space Applications. *Compos. Sci. Technol.* **2005**, *65*, 741–748. [CrossRef]

28. Voevodin, A.A.; O'neill, J.P.; Zabinski, J.S. Nanocomposite Tribological Coatings for Aerospace Applications. *Surf. Coat. Technol.* **1999**, *116–119*, 36–45. [CrossRef]
29. Duan, Z.; Zhao, X.; Nai, Z.; Qiao, L.; Xu, J.; Wang, P.; Liu, W. Mo-S-Ti-C Nanocomposite Films for Solid-State Lubrication. *ACS Appl. Nano Mater.* **2019**, *2*, 1302–1312. [CrossRef]
30. Bülbül, F.; Efeoğlu, İ. Synergistic Effect of Bias and Target Currents for Magnetron Sputtered MoS2-Ti Composite Films. *Mater. Test.* **2016**, *58*, 471–474. [CrossRef]
31. Chien, H.; Ma, K.; Vattikuti, S.V.P.; Kuo, C.; Huo, C.; Chao, C. Tribological Behaviour of MoS2/Au Coatings. *Thin Solid Films* **2010**, *518*, 7532–7534. [CrossRef]
32. Scharf, T.W.; Goeke, R.S.; Kotula, P.G.; Prasad, S.V. Synthesis of Au-MoS2 Nanocomposites: Thermal and Friction-Induced Changes to the Structure. *ACS Appl. Mater. Interfaces* **2013**, *5*, 11762–11767. [CrossRef]
33. Zekonyte, J.; Cavaleiro, A.; Polcar, T. Frictional Properties of Self-Adaptive Chromium Doped Tungsten-Sulfur- Carbon Coatings at Nanoscale. *Appl. Surf. Sci.* **2014**, *303*, 381–387. [CrossRef]
34. Zekonyte, J.; Polcar, T. Friction Force Microscopy Analysis of Self-Adaptive W—S—C Coatings: Nanoscale Friction and Wear. *ACS Appl. Mater. Interfaces* **2015**, *7*, 21056–21064. [CrossRef]
35. Cao, H.; De Hosson, J.T.M.; Pei, Y. Effect of Carbon Concentration and Argon Flow Rate on the Microstructure and Triboperformance of Magnetron Sputtered WS2/a-C Coatings. *Surf. Coat. Technol.* **2017**, *332*, 142–152. [CrossRef]
36. Isaeva, L.; Sundberg, J.; Mukherjee, S.; Pelliccione, C.J.; Lindblad, A.; Segre, C.U.; Jansson, U.; Sarma, D.D.; Eriksson, O.; Kádas, K. Amorphous W-S-N Thin Films: The Atomic Structure behind Ultra-Low Friction. *Acta Mater.* **2015**, *82*, 84–93. [CrossRef]
37. Seynstahl, A.; Köbrich, M.; Rosnitschek, T.; Göken, M.; Tremmel, S. Enhancing the Lifetime and Vacuum Tribological Performance of PVD-MoS2 Coatings by Nitrogen Modification. *Surf. Coat. Technol.* **2024**, *477*, 130343. [CrossRef]
38. Yaqub, T.B.; Fernandes, F.; Al-Rjoub, A.; Cavaleiro, A. Mo-Se-N Dry Lubricant Coatings as a Universal Solution for Protecting Surfaces of Complex 3D Parts. *Mater. Lett.* **2022**, *316*, 131967. [CrossRef]
39. Gustavsson, F.; Jacobson, S.; Cavaleiro, A.; Polcar, T. Frictional Behavior of Self-Adaptive Nanostructural Mo-Se-C Coatings in Different Sliding Conditions. *Wear* **2013**, *303*, 286–296. [CrossRef]
40. Gao, J.; Li, B.; Tan, J.; Chow, P.; Lu, T.M.; Koratkar, N. Aging of Transition Metal Dichalcogenide Monolayers. *ACS Nano* **2016**, *10*, 2628–2635. [CrossRef]
41. Liu, H.; Han, N.; Zhao, J. Atomistic Insight into the Oxidation of Monolayer Transition Metal Dichalcogenides: From Structures to Electronic Properties. *RSC Adv.* **2015**, *5*, 17572–17581. [CrossRef]
42. Rajput, N.S.; Kotbi, A.; Kaja, K.; Jouiad, M. Long-Term Aging of CVD Grown 2D-MoS2 Nanosheets in Ambient Environment. *Npj Mater. Degrad.* **2022**, *6*, 75. [CrossRef]
43. Vuchkov, T.; Evaristo, M.; Yaqub, T.B.; Cavaleiro, A. The Effect of Substrate Location on the Composition, Microstructure and Mechano-Tribological Properties of W-S-C Coatings Deposited by Magnetron Sputtering. *Surf. Coat. Technol.* **2020**, *386*, 125481. [CrossRef]
44. Voevodin, A.A.; O'Neill, J.P.; Prasad, S.V.; Zabinski, J.S. Nanocrystalline WC and WC/a-C Composite Coatings Produced from Intersected Plasma Fluxes at Low Deposition Temperatures. *J. Vac. Sci. Technol. A Vac. Surf. Film.* **1999**, *17*, 986–992. [CrossRef]
45. Polcar, T.; Cavaleiro, A. Review on Self-Lubricant Transition Metal Dichalcogenide Nanocomposite Coatings Alloyed with Carbon. *Surf. Coat. Technol.* **2011**, *206*, 686–695. [CrossRef]
46. Yaqub, T.B.; Vuchkov, T.; Bruyère, S.; Pierson, J.F.; Cavaleiro, A. A Revised Interpretation of the Mechanisms Governing Low Friction Tribolayer Formation in Alloyed-TMD Self-Lubricating Coatings. *Appl. Surf. Sci.* **2022**, *571*, 151302. [CrossRef]
47. Yaqub, T.B.; Bruyere, S.; Pierson, J.F.; Vuchkov, T.; Cavaleiro, A. Insights into the Wear Track Evolution with Sliding Cycles of Carbon-Alloyed Transition Metal Dichalcogenide Coatings. *Surf. Coat. Technol.* **2020**, *403*, 126360. [CrossRef]

nanomaterials

MDPI

Article

Impact of Ni Doping on the Microstructure and Mechanical Properties of TiB_2 Films

Ying Wang [1], Xu Wang [1], Hailong Shang [1], Xiaotong Liu [1], Yu Qi [1], Xiaoben Qi [1,*] and Ning Zhong [2]

[1] School of Materials Science, Shanghai Dianji University, Shanghai 201306, China; ywang@sdju.edu.cn (Y.W.); wx19980927@163.com (X.W.); shanghl@sdju.edu.cn (H.S.); dndxj@126.com (X.L.); m15135239996@163.com (Y.Q.)

[2] College of Ocean Science and Engineering, Shanghai Maritime University, Shanghai 201306, China; ningzhong@shmtu.edu.cn

* Correspondence: qixb@sdju.edu.cn

Abstract: The TiB_2 film exhibits exceptional hardness and chemical stability due to its unique crystal structure and robust covalent bonds, but it also demonstrates high brittleness and poor toughness, which restricts its practical applications in engineering. By appropriately incorporating metal dopants, the toughness of the ceramic matrix can be enhanced without compromising its inherent hardness. In this study, TiB_2 films with different nickel contents (0–32.22 at.%) were fabricated through radio frequency magnetron sputtering. The microstructure, chemical composition, phase structure, and mechanical properties were analyzed using scanning electron microscopy, transmission electron microscopy, X-ray diffraction, X-ray photoelectron spectroscopy and nanoindentation tester. The pure TiB_2 film exhibited (0001) and (0002) peaks; however, the addition of nickel resulted in broadening of the (0001) peak and disappearance of the (0002) peak, and no crystalline nickel or other nickel-containing phases could be identified. It was found that the incorporation of nickel refines the grain structure of titanium diboride, with nickel present in an amorphous form at the boundaries of titanium diboride, thereby forming a wrapped structure. The enrichment of nickel at the grain boundary becomes more pronounced as the nickel content is further increased, which hinders the growth of TiB_2 grains, resulting in the thinning of columnar crystals and formation of nanocrystalline in the film, and the coating hardness remains above 20 GPa, when the nickel content is less than 10.83 at.%. With the increase in nickel content, titanium diboride exhibited a tendency to form an amorphous structure, while nickel became increasingly enriched at the boundaries, and the coating hardness and elastic modulus decreased. The wrapped microstructure could absorb the energy generated by compressive shear stress through plastic deformation, which should be beneficial to improve the toughness of the coatings. The addition of nickel enhanced the adhesion between the film and substrate while reducing the friction coefficient of the film. Specifically, when the nickel content reached 4.26 at.%, a notable enhancement in both nanohardness and toughness was observed for nanocomposite films.

Keywords: TiB_2 film; Ni doping; nanocomposite; microstructure; mechanical property

1. Introduction

Due to the unique chemical bond composition of TiB_2 ceramics, they exhibit exceptionally high hardness, melting point, and excellent chemical stability [1,2]. As a result, new TiB_2-based coatings have been extensively developed. Various coating preparation technologies exist including laser cladding [3], thermal spraying [4], electrochemical deposition [5], chemical vapor deposition [6], physical vapor deposition [7], and more. Among these, sputtering technology is commonly used due to its low deposition temperature requirement and relatively high coating rates [8]. Researchers [9,10] have discovered that boron-rich TiB_x films with x typically ranging from 2.2 to 3.5 can be obtained using magnetron sputtering. Studies by Christian Mitterer [11] suggested that B atoms in the sputtered plasma have a smaller collision cross-sectional area and energy transfer coefficient compared to Ti atoms. This results in the B atoms being less likely to collide with Ar atoms during the plasma transport process leading to lower power loss after collisions and increased likelihood of B atoms being deposited on the substrate surface. In addition, high-power impact magnetron sputtering has been employed to produce highly understoichiometric TiB_x with x as low as ~1.4 [12]. Researchers [13] propose that excess Ti leads to planar stacking fault defects with islands where one or two boron planes are missing in the films. The disruption of crystalline structures inhibits dislocation propagation affecting mechanical properties. Moreover, it has been reported by numerous researchers that the hardness of the TiB_2 coating is closely related to optimal orientation which can be regulated by parameters of magnetron sputtering. It is often observed that the high hardness of the TiB_2 coating exhibits a strong (0001) preferred orientation [14,15].

However, the main drawback of TiB_2 coatings is the low toughness, high residual stress, and low cracking resistance in the deposited films [16,17], making it difficult to achieve the desired mechanical properties. In addition to optimizing coating properties by adjusting deposition parameters, two other effective methods for improving the performance of hard films are through multilayer structures [18,19] and nanocomposites [20]. The structural coating with a periodic arrangement is achieved through alternating deposition, and its properties are influenced by both the multilayer period and the interface between adjacent layers. Douglas E. Wolfe [21] found that the stress of TiB_2/TiC multilayers decreases with an increase in the number of layers, thereby enhancing the film toughness. O.N. Grigor'ev [22] prepared a TiB_2 coating containing a β-SiC layer with a porous arch structure which eliminates crack formation due to its high relaxation capacity. The intricate influence of interface structure and interlayer in multilayer films can impede dislocation slip and crack propagation [23], thereby not only contributing to toughening but also enhancing other properties.

Element doping is an additional effective approach to modify the properties of the film, thereby exerting influence on its structure and phase composition, consequently impacting both hardness and toughness. The coexistence of the doped second phase is facilitated by the narrow phase interval of TiB_2. The findings of these studies by adding Cu [24], Ni [25], Cr [26], C [27], W [28], Al [5], Zr [29] and high-entropy alloy elements [30] to form a nanocomposite structure demonstrate that the incorporation of dopant elements can effectively restrict the growth of TiB_2 grains, leading to a reduction in grain size. The study conducted by Nedfors et al. [31] revealed that the incorporation of aluminum into the TiB_2 coating resulted in an improved ratio of hardness to Young's modulus, thereby enhancing its toughness, The presence of the second phase in nanocrystalline or amorphous form enables effective control over the plastic zone at the crack tip, impeding crack propagation and enhancing the coating's toughness. The influence of adding Al, Ni, and Zr on the crystal structure and elastic constant of TiB_2 coatings was theoretically analyzed using first principles by Xian Lijun [32], and the results indicated that when Ti atoms in the lattice of TiB_2 are partially replaced by Al, Ni and Zr atoms, the cell volume and the binding energy of the $(Ti_{1-x}M_x)B_2$ ternary coatings increases, which leads to a reduction in the hardness and the improvement of the coating toughness, and the addition of nickel has a more pronounced impact on the reduction in hardness of the TiB_2 coating.

Due to the good wettability of nickel and TiB_2, as well as the rapid diffusion rate of nickel within the hexagonal lattice [33,34], favorable conditions are created for the formation of TiB_2-Ni nanocomposite films. The study of obtaining a film with both toughness and hardness is worth pursuing. In this study, TiB_2 films with different nickel concentrations were deposited using radio frequency magnetron sputtering. The microstructure, chemical and phase composition, as well as the mechanical properties of Ni-doped TiB_2 films, were characterized using X-ray diffraction, X-ray photoelectron spectroscopy, scanning electron microscopy with energy-dispersive X-ray spectroscopy, transmission electron microscopy, nanoindentation techniques and Tribology Test Equipment. Subsequently, the effects of different Ni content on the microstructure, nanohardness, toughness and wear performance of TiB_2 coatings were systematically evaluated.

2. Materials and Methods

2.1. Film Deposition

TiB_2 films with different nickel content were deposited onto single-crystal silicon substrates (10 mm × 10 mm × 1 mm) and using magnetron sputtering (ANAVA SPC-350, Anelva, Tokyo, Japan). A twin radio frequency current magnetron setup, equipped with disk-shaped targets made of pure nickel (Ni, ϕ 76 mm) and titanium diboride (TiB_2, ϕ 76 mm), was employed for the deposition. After ultrasonic cleaning with acetone for 10 min, the substrate was subjected to constant temperature drying, followed by placement into the vacuum chamber and subsequent vacuuming to achieve a pressure of 3×10^{-1} Pa.

Radio frequency magnetron sputtering is particularly suitable for materials with poor electrical conductivity or insulating properties, such as ceramics and glass. Sputtering power, argon pressure, substrate rotation speed and substrate temperature are critical parameters that significantly influence the sputtering process. The sputtering power dictates the energy imparted to the target material, thereby influencing both the quantity and energy distribution of the sputtered particles. With the increase in power, the deposition rate increases, and when the power is excessively high, it can lead to the sputtered atoms arriving at the substrate surface with elevated energy levels. It induces significant internal stress within the film, thereby reducing the adhesion between the film and the substrate. Higher pressure typically results in greater plasma density, but it can also increase the frequency of collisions between sputtered particles, which may compromise the uniformity and quality of the deposited film. Conversely, lower air pressure facilitates the formation of high-energy sputtered particles, making it more suitable for applications that demand highly dense and uniform films. The substrate temperature plays a crucial role in determining the film structure, grain growth and binding strength. High temperature facilitates crystal growth and densification of the film, but it may also induce thermal deformation or damage to the substrate. It can be controlled directly through the substrate heating system or indirectly by adjusting the sputtering power and chamber pressure.

In this paper, the argon pressure in the vacuum chamber during sputtering was maintained at 0.5 Pa, while the power on the Ni target was adjusted within the range of 5 to 30 W in order to achieve different Ni contents. The power on the TiB_2 target remained constant at 180 W throughout. The substrate rotated at a speed of 20 revolutions per minute during deposition, and each film was sputtered for a duration of 4 h. The composite sputtering deposition was achieved through the cyclic rotation of the substrate frame in front of the target, enabling the adjustment of target power to obtain TiB_2-Ni films with varying amounts of Ni doping.

2.2. Film Characterization

Phase analyses of the films were conducted using X-ray diffraction with a Bruker D8 ADVANCE diffractometer (Brucker, Karlsruhe, Germany) equipped with a Cu K_{α} lamp (0.154 nm). The cross-sectional morphology of the films after sputtering was observed using scanning electron microscopy (ZEISS Sigma300, ZEISS, Oberkochen, Germany) and transmission electron microscopy (JE-2010ARP, JEOL, Tokyo, Japan) by preparing cross-sectional lamella manufactured via focused ion beam (FIB). Selected area electron diffraction patterns, obtained using an ϕ 150 nm aperture, were used for phase analysis.

Additionally, the composition of the films was analyzed using an energy-dispersive spectrometer (Oxford Xplore30, Oxford Instruments, Oxford, UK). X-ray photoelectron spectroscopy (XPS) with a Thermo Fisher Scientific K-Alpha instrument (Thermo Fisher Scientific, Waltham, United States) was utilized for qualitative and quantitative analysis of the chemical bond state and phase structure of elements in the films. The Al K_{α} source with the energy of 1486.6 eV was used, and the spot size was 400 μm.

The nanoindentation technique was employed to investigate the nanohardness of TiB_2 films with different Ni contents, using Paar's Step300-NTH3 instrument (Anton Paar, Graz, Austria) equipped with a Berkovich diamond tip. A maximum force of 10 mN was applied, with loading and unloading rates set at 30 mN/min. The values of hardness (H), elastic modulus (E), maximum indentation depth (h_{max}), and indentation depth (h_p) were analyzed through the load–displacement curve using the Oliver–Pharr method. Parameters such as H/E, H^3/E^2, and normalized plastic depth value (δ_H) were calculated to evaluate the mechanical properties of the films, where the normalized plastic depth value is equivalent to the indentation depth (h_p) divided by maximum indentation depth (h_{max}). At least nine measurements were conducted for each load condition.

The fracture toughness of the films was determined using the Vickers indentation method, with a load of 500 mN applied to the film surface. After unloading, the fracture toughness can be calculated according to Equation (1) [35] by examining the indentation morphology and measuring the diagonal crack length of the indentation.

$$K_{IC} = \delta\left(\frac{a}{l}\right)^{\frac{1}{2}}\left(\frac{E}{H}\right)^{\frac{2}{3}}\left(\frac{P}{c^{3/2}}\right) \quad (1)$$

where E (GPa) and H (GPa) represent the elastic modulus and hardness of the films, respectively. P (mN) stands for Vickers indentation load. The empirical coefficient (δ) is 0.016 [36], and c (μm) denotes the total length from the center to the end of crack as shown in Figure 1, where a represents half of the diagonal length of the indentation, and l represents the average length of cracks. The indentation morphology of films was observed using a microscope (Leica DCM 3D, Wetzlar, Germany).

The nanoscratch tests were performed on Tribology Test Equipment (CETR UNMT-1, CETR, Campbell, United States). The diamond tip with the radius of 2.5 μm was set to generate a series of 2 mm-long wear tracks on film surface at a constant velocity of 12.5 μm/s, with the load increasing linearly from 2 mN to 50 mN. The scratch morphology was observed using a microscope (Leica DCM 3D, Wetzlar, Germany).

The wear resistance tests were performed on Tribology Test Equipment (CETR UNMT-1, CETR, Campbell, CA, USA). At room temperature, a reciprocating wear test was conducted using an Al_2O_3 grinding ball with a diameter of 1.5 mm. The applied load was 20 mN, the reciprocating friction distance was 1 cm, and the duration time was 15 min. The surface friction coefficient was recorded, while the wear tracks were observed using scanning electron microscopy (S-3400N, Hitachi, Tokyo, Japan).

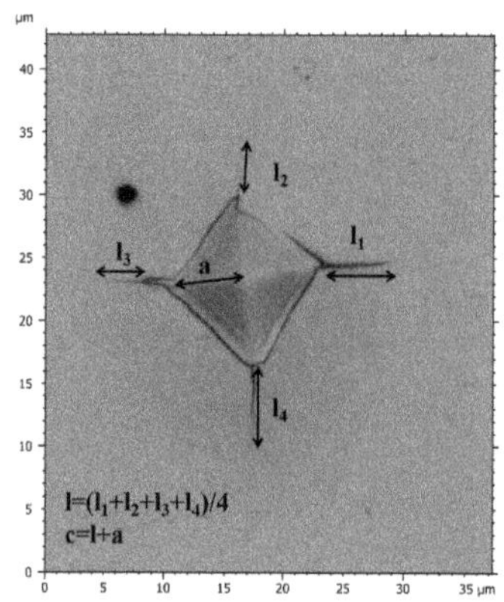

Figure 1. Schematic diagram of crack length (c) and half of the diagonal length of the indentation (a).

3. Results and Discussion

3.1. Effect of Ni on the Phase Structure

TiB_2 films with different nickel contents were deposited using radio frequency magnetron sputtering. The film composition obtained using EDS analysis are shown in Table 1. As the Ni target power increased, the Ni content ranged from 0 to 32.2 at.%. The increase in power applied to the nickel target leads to the back-sputtering of the lighter B element from the substrate film, resulting in a relative decrease in the B element within the coating.

Table 1. The element composition of TiB_2 films with different Ni contents.

The Power on Ni Target/W	At.%			Ni/Ti
	Ni	Ti	B	
0	-	22.20	77.80	-
5	4.26	22.23	73.51	0.19
10	10.83	25.23	63.94	0.43
20	23.45	22.75	53.80	1.03
30	32.22	19.53	48.25	1.65

In Figure 2, the X-ray diffraction patterns of TiB_2 films with different Ni contents are presented. The findings indicate that all films were polycrystalline, with peaks corresponding exclusively to the hexagonal TiB_2 phase (PDF#35-0741), and no crystalline Ni or other Ni-containing phases were identified. In comparison to the standard card of hexagonal TiB_2, diffraction peaks of (0001) and (0002) were observed. It was noted that when nickel was incorporated into the films, the diffraction peak of (0002) disappeared. The growth of the film is influenced by surface energy, elastic strain energy, and interfacial energy. When the dominant factor is surface energy, the film preferentially grows along crystal faces with lower surface energy, and the (0001) crystal face exhibits the lowest surface energy for TiB_2 film. The incorporation of nickel (4.26–23.45 at.%) did not influence the favored alignment of the TiB_2 coating, but there was a gradual decrease in intensity and broadening of the (0001) diffraction peaks of TiB_2 phase. When the nickel content increased to 32.22%, the diffraction pattern became nearly linear, indicating that the addition of Ni impeded the growth of TiB_2 grains and resulted in a reduction in their size and an amorphous tendency. The results indicate that the presence of nickel hinders the nucleation and growth of TiB_2 grains, and this hindrance becomes more pronounced as the nickel content increases. The faster diffusion rate of nickel atoms in the hexagonal lattice may account for this phenomenon. As the nickel content increases, an increasing number of nickel atoms are enriched in the grain boundary, which hinders the growth of grain. No crystalline Ni or other Ni-containing phase could be identified, indicating that nickel was present in an amorphous phase. Overall, these results provide valuable insights regarding how varying levels of nickel content can impact structural characteristics within TiB_2 films.

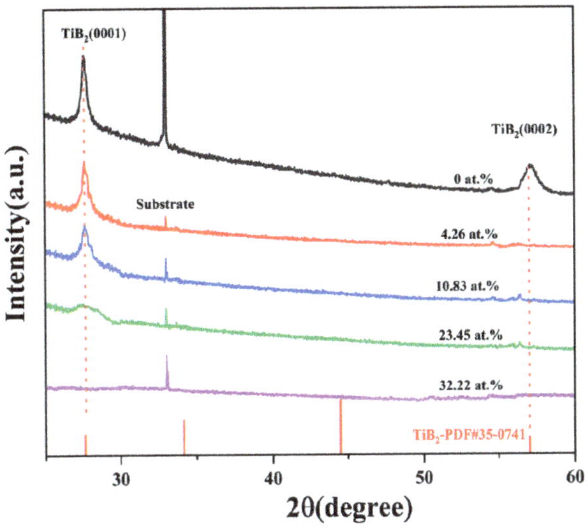

Figure 2. X-ray diffraction patterns of TiB_2 films with different Ni contents.

The deposited TiB_2 coating was subjected to X-ray photoelectron spectroscopy analysis in order to determine the chemical elements present and their respective bonding states. Firstly, the surfaces of spectral lines of B 1s, Ni 2p and Ti 2p were analyzed, and the results of the TiB_2-Ni (4.26 at.%) film are presented in Figure 3. The spectra were calibrated for the value of carbon peak C 1 s at 284.8 eV, which was used as a reference. In Figure 3a, the B-Ti bond was observed at 187.85 eV, corresponding to the presence of TiB_2. Figure 3b depicts the Ni 2p spectrum of the TiB_2 Ni film, and the Ni-Ni bond was identified at the binding energies of 869.84 eV and 852.67 eV. The Ti 2p region is considerably more complex due to the spin–orbit splitting and higher electron yield of titanium, which results in a higher background step and more prominent plasmon peaks [37]. As given in Figure 3c, it can be seen that the Ti-B bond can be fitted at 454.51 eV and 460.67 eV by dividing the peaks of $Ti2p_{2/3}$ and $Ti2p_{1/2}$, which correspond to TiB_2. In the meantime, the Ti-Ni peak cannot be fitted from Ti2p spectra, and no peak for B-Ni bonding could be fitted from the B 1s spectrum.

Additionally, the B-O bond peak was detected at the position of 192.48 eV in the B 1S map (Figure 3a), and the Ti-O bond peak was also detected in the Ti-2p map, which indicates that element boron and titanium samples are usually oxidized at the surface. To avoid the impact of surface contaminants on the results, the sample surface was etched using Ar+ for 120 s, and Figure 4 shows the XPS spectra of TiB_2-Ni films with different nickel content. As shown in Figure 4a, the B-O bond peak was not detected again; moreover, the oxygen-containing bond in the coatings was significantly weakened. It also can be seen that the valence state of elements in the coatings does not change significantly with the addition of nickel content.

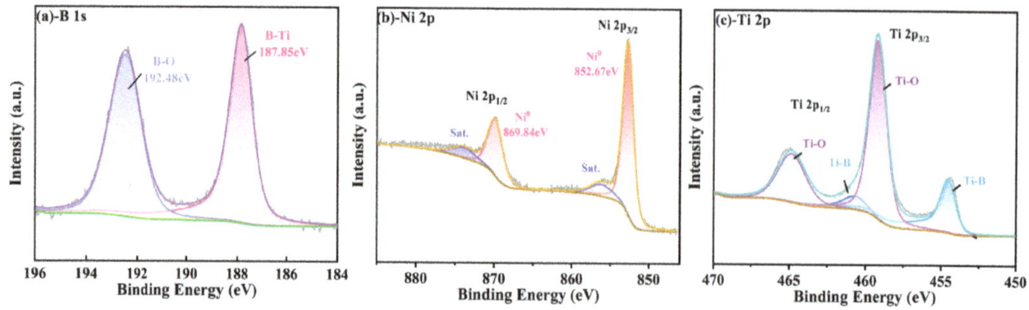

Figure 3. XPS spectra of B, Ni Ti, obtained from the surface of TiB_2 film with 4.26 at.% Ni (**a**) -B 1s; (**b**) -Ni 2p; (c) -Ti 2p.

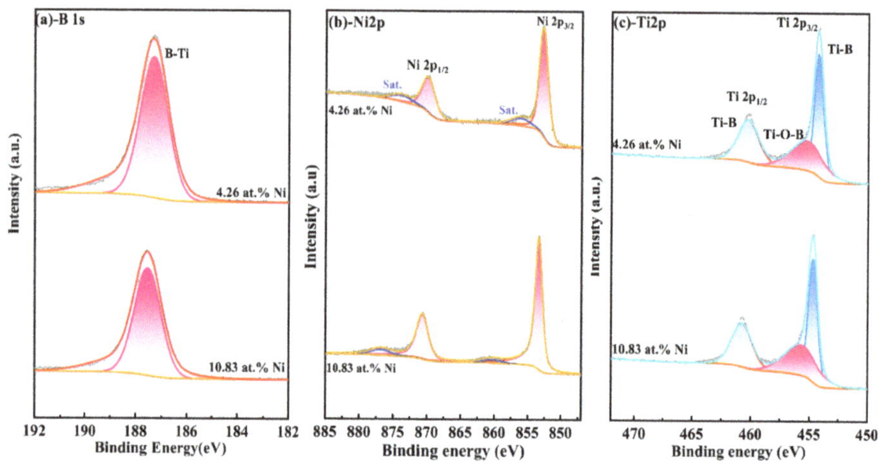

Figure 4. XPS spectra of TiB_2-Ni films with different nickel content after etching 120 s (**a**) -B 1s; (**b**) -Ni 2p; (c) -Ti 2p.

The findings indicate that there is no chemical bonding between Ni and Ti/B atoms within the film, which demonstrates that the main chemical form of the Ni is the metallic state. And the presence of Ti-O and B-O in XPS results may be related to the residual oxygen in the vacuum chamber, since no oxide phases of Ti and B were found in XRD spectra.

Figure 5 depicts a cross-sectional SEM micrograph of a TiB_2 film with varying Ni contents. It is evident from Figure 5 that there is an absence of any voids or gaps between the film and substrate, which indicates a good adhesion between the coating and substrate. The columnar crystals exhibit a typical growth structure when the nickel content is 4.26%, as shown in Figure 5a. With the increase in nickel content, the columnar crystals within the coating are gradually eliminated, the fine grains undergo deformation, and the overall columnar structure is progressively refined.

The transmission electron microscopy (TEM) results for TiB_2 films with nickel content of 4.26 at.% and 32.22 at.% are presented in Figure 6. The bright-field cross-sectional images of the films are shown in Figure 6a,d, revealing a predominant presence of small columnar grains in the film microstructure. At a nickel content of 4.26 at.%, the selected area diffraction pattern (Figure 6b) exhibits a polycrystalline diffraction ring, and the reflection peaks such as (0001) are observed without any evidence of crystalline Ni phase, suggesting the existence of a small amount of TiB_2 nanocrystals and amorphous nickel within the film structure. With an increase in nickel content to 32.22 at.%, there is an observable broadening of the electron diffraction ring (Figure 6e), indicating that the composite film is predominantly amorphous. The combination of the local high-resolution image (Figure 6c,f) reveals a homogeneous structure in the film, wherein the addition of nickel hinders the growth of TiB_2 grains, resulting in a tendency towards amorphousness. The microstructure of TiB_2-Ni films was significantly influenced by the Ni content. Increasing the Ni content led to grain refinement and a more compact and denser film structure. It should be noted that no crystalline Ni phase was observed in any of the films, regardless of their Ni contents.

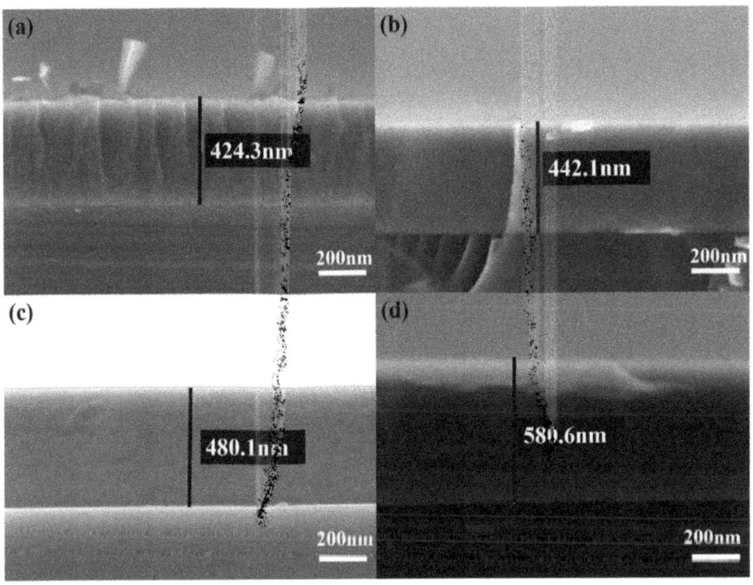

Figure 5. Cross-sectional SEM micrographs of TiB_2 film with different Ni contents: (**a**) 4.26 at.%; (**b**) 10.83 at.%; (**c**) 23.45 at.%; (**d**) 32.22 at.%.

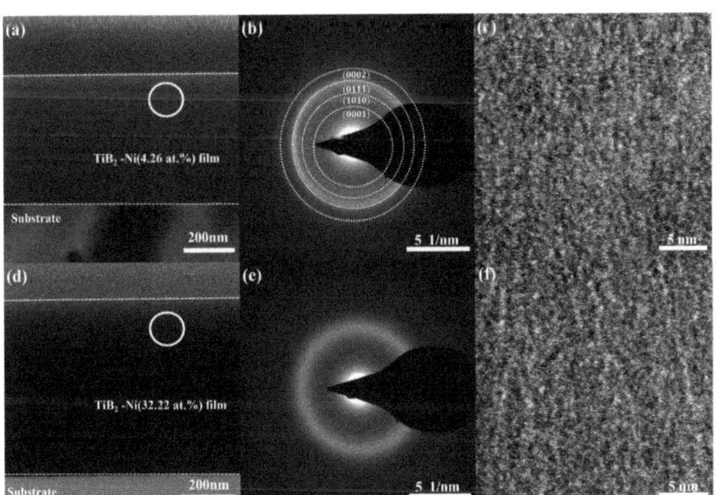

Figure 6. TEM results for TiB_2 films with different Ni contents: (**a**) bright-field cross-sectional image (4.26 at.%); (**b**) SAED pattern from the area marked as circle in (**a**); (**c**) HRTEM image; (**d**) bright-field cross-sectional image (32.22 at.%); (**e**) SAED pattern from the area marked as circle in (**d**); (**f**) HRTEM image.

Figure 7 presents the high-angle annual dark-field (HAADF) image of TiB_2-Ni (4.26 and 32.22 at.%), along with the element mapping scans for B, Ti and Ni. As depicted in Figure 7b–d, the distribution of nickel in the TiB_2 coating is uniform with a nickel content of 4.26 at.% (Figure 7a). When the nickel content increased to 32.22 at.% (Figure 7d), the distribution of B element became limited. With the increase in the nickel target power, the more B atoms on the surface of the substrate are sputtered, resulting in a decrease in element B, and the amorphous-phase Ni atoms can be randomly dispersed within the coatings.

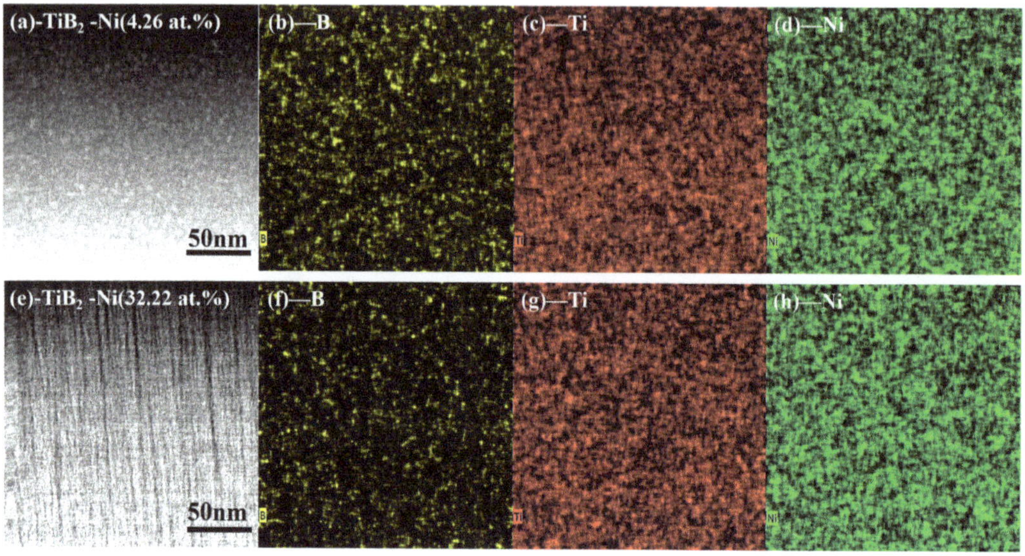

Figure 7. HAADF images and the corresponding EDX map scan of B, Ti and Ni elements of TiB_2-Ni films. (**a**) Image of TiB_2 with 4.26 at.% Ni film; (**b**) map scan of B; (**c**) map scan of Ti; (**d**) map scan of Ni; (**e**) Image of TiB_2 with 32.22 at.% Ni film; (**f**) map scan of B; (**g**) map scan of Ti; (**h**) map scan of Ni.

The TEM observations combined with the XPS results and XRD results verified that the Ni atoms existed in films mainly as an amorphous phase. To further investigate the relationship between nickel and titanium diboride in the coatings, the surface of the composite coating was characterized using transmission electron microscopy. Figure 8 shows the plan-view TEM micrographs of TiB_2 with 10.83 at.% Ni coating. It can be seen from the figure that the coating has a two-phase structure with white bright bands surrounding dark particles (Figure 8a). Further analysis (Figure 8b) reveals that the dark particles are TiB_2 nanocrystals with a grain size ranging from 3 to 5 nm, and nickel is present as an amorphous phase at the boundaries of the TiB_2. The addition of nickel exists as a single phase, contributing to the development of a two-phase composite structure. The diffusion rate of nickel in the hexagonal lattice is significantly higher [33], facilitating its migration to the grain boundary of TiB_2, thereby effectively impeding the grain growth of TiB_2.

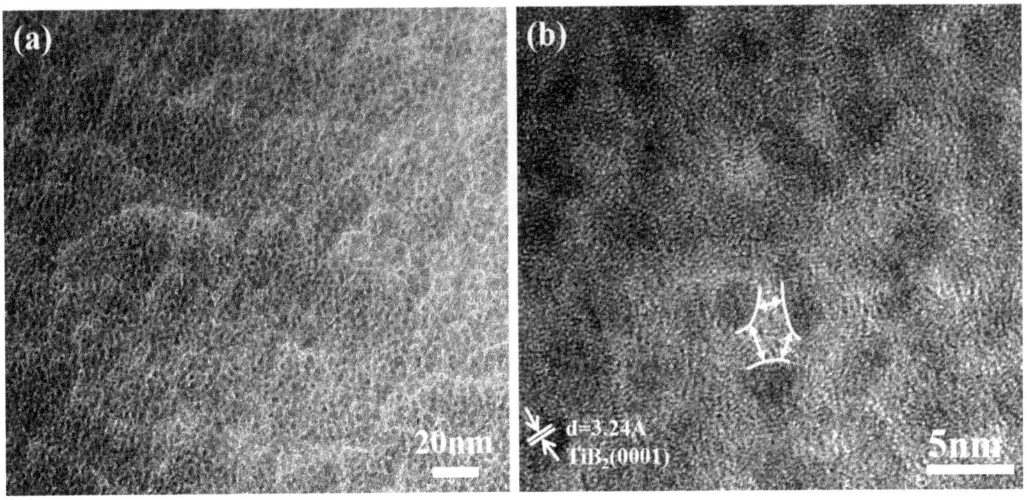

Figure 8. Plan-view TEM micrograph of TiB_2 with 10.83 at.% Ni coating. (**a**) low-magnification; (**b**) high-magnification.

3.2. Effect of Ni on Mechanical Properties

Figure 9 illustrates the load–indentation-depth curves of TiB_2 films with varying Ni contents. The loading and unloading curves of the films exhibit nonlinearity. Under the same load, the maximum indentation depth and the residual indentation depth gradually increase with the increase in nickel content. As shown in Figure 10, the results indicate that there is an initial increase followed by a subsequent decrease in nanohardness. The nanohardness of the TiB_2 film is determined to be 31 $\pm$ 2.2 GPa. With the addition of nickel of 4.26 at.%, the nanohardness of the composite film reaches 34.8 $\pm$ 1.2 GPa. However, as the nickel content further increases to 32.22 at.%, a continuous decrease in nanohardness to 17.89 $\pm$ 1.8 GPa is observed for the composite film. As depicted in Figure 10, the curve of elastic modulus exhibits a similar trend to that of nanohardness. Specifically, the elastic modulus measures 271 $\pm$ 15 GPa, at a Ni content of 4.26 at.%. As the nickel content further increases, the elastic modulus of the film gradually decreases, reaching 230 $\pm$ 9 GPa at a nickel content of 32.22 at.%.

When the nickel content is below 10.83 at.%, the coating hardness remains at or above 20 GPa. On one hand, the optimal growth of TiB_2 (0001) crystal surface helps improve the film hardness [34]. On the other hand, it may be attributed to the grain refinement effect of TiB_2 and the inhibition of grain boundary sliding by the nickel amorphous phase [38]. The theoretical analysis in reference [32] also indicates that when the nickel content is below 13 at.%, the hardness of the coating exhibits a slight decrease; however, it remains comparable to that of the pure titanium diboride coating.

As the nickel content progressively increases, the hardness of the coatings decreases. An increase in nickel content results in a notable reduction in boron (B) concentration within the film, consequently diminishing the number of Ti-B and B-B bonds. The enrichment of nickel in the boundaries of titanium diboride increased with the increase in nickel element. Additionally, it is evident from the XRD pattern that the addition of nickel leads to a reduction in the crystallinity of titanium diboride, consequently altering the hardness of the coatings. Consequently, an increase in Ni content results in an increased load for the softer Ni phase to bear, consequently leading to a decrease in hardness for the composite film as well.

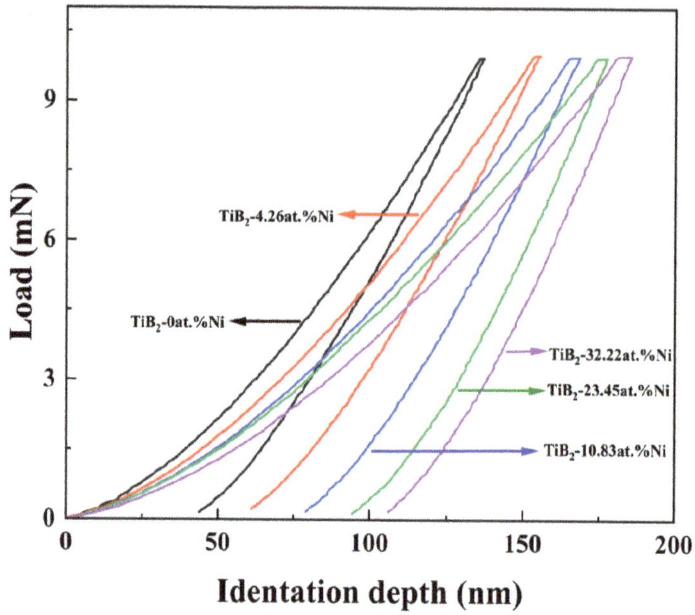

Figure 9. The indentation curves of TiB_2 film with different Ni contents.

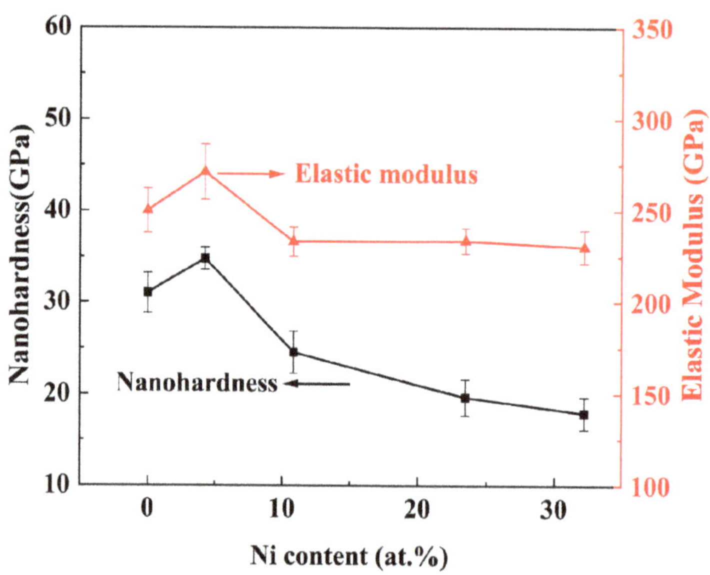

Figure 10. The nanohardness and elastic modulus of TiB_2 film with different Ni contents.

With regard to the elastic behavior, it is not sufficient to describe the film properties solely based on hardness; instead, the ratio of hardness (H) to elastic modulus (E) should be considered [39]. The H/E ratio indicates the resistance to cracking, while the H^3/E^2 ratio reflects the resistance to plastic deformation [40]. Figure 11 illustrates the values of H/E and H^3/E^2 for TiB_2 films with varying Ni contents. It can be observed that when the nickel content is 4.26 at.%, the values of H/E and H^3/E^2 reach their maximum at 0.128 and 0.573 GPa, respectively. When nickel is dissolved into the lattice of titanium diboride, it causes lattice distortion, resulting in an increase in the internal stress of the coating and an increase in the elastic modulus [41]. Previous studies [16,42] have indicated that a high value of H^3/E^2 along with a high ratio of H/E (>0.1), combined with a dense microstructure, are favorable for enhancing film toughness. However, when the nickel content exceeds 10.83%, the value of H/E drops below 0.1. Upon increasing nickel content from 10.83 at.% to 32.22 at.%, both ratios gradually decrease. The two values of H/E and H^3/E^2 reflect key parameters relating to film resistance against elastoplastic deformation, and demonstrate positive correlations with fracture toughness as well as wear resistance [43].

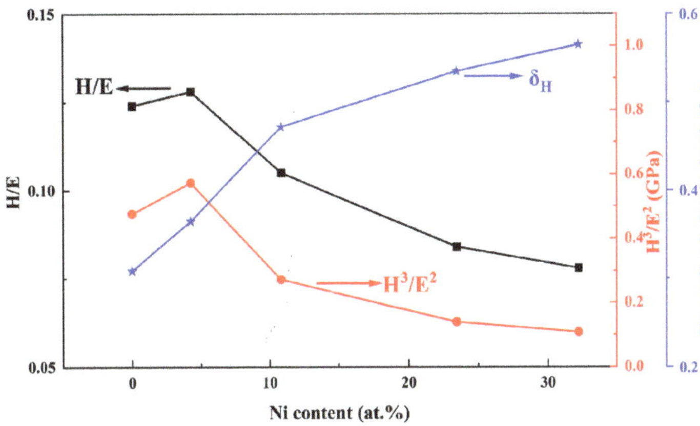

Figure 11. The values of H/E, H^3/E^2, normalized plastic depth value (δ_H) of TiB_2 film with different Ni contents.

The normalized plastic depth value (δ_H) is also utilized to assess the film's plasticity, as illustrated in Figure 11. The plastic deformation capacity [44] serves as an indicator of toughness to a certain extent, with higher toughness associated with stronger plastic deformation capacity. It can be observed from Figure 10 that with an increase in Ni content from 0 at.% to 10.83 at.%, there is a large rise in the normalized plastic depth. As the content continues to increase, the rate of growth in the normalized plastic depth decelerates. With an increase in the depth of deformation, there is a corresponding enhancement in plasticity index, indicating greater plastic deformation ability and improved toughness.

The nanoindentation test has proven to be a powerful technique to evaluate material's brittleness or resistance to fracture [45]. The calculated results of fracture toughness of the films with different nickel contents used by Equation (1) are given in Figure 12. As illustrated in the figure, when the nickel content increases to 10.83 at.%, the toughness of the coatings increases faster. The TiB_2 film exhibits a fracture toughness of 0.69 $MPa\cdot m^{1/2}$, and this value increases to 0.91 $MPa\cdot m^{1/2}$ when the nickel content reaches 10.83 at.%. When the nickel content increases further, the film toughness slows down, and the value increases to 1.06 $MPa\cdot m^{1/2}$ when the nickel content reaches 32.22 at.% in composite films. This is consistent with the trend of normalized plastic depth.

Figure 13 displays the indentation morphology of the TiB_2 film with varying Ni contents (4.26 at.% and 32.22 at.%). It is evident that circular cracks appear around the indentation, indicating a tendency toward brittle fractures. The formation and propagation of these circular cracks are primarily attributed to macroscopic plastic deformation, which leads to crack propagation during continuous loading due to uncoordinated deformation between the film and substrate. However, as the Ni content increases, the presence and visibility of circular cracks decrease gradually under identical load conditions, suggesting an increase in the fracture toughness of films as well. Toughness refers to the film's ability to absorb energy between deformation and fracture [46]. The increase in nickel content leads to a reduction in circumferential cracks of the coating under the same load, gradually diminishing until they disappear completely. Simultaneously, based on the three-dimensional indentation morphology (Figure 12), it is evident that an increase in nickel content leads to a more homogeneous deformation of the film. When the nickel content reached 4.26 at.%, annular cracks manifested both around and within the indentation, accompanied by non-uniform deformation. Under the same load, annular cracks did not appear inside the indentation, and the film deformation was uniform when the nickel content increased to 32.22 at.%. This observation suggests an enhancement in the toughness of the coating with the increase in nickel content. When crack propagation occurs in the films, the Ni metal phase exhibits excellent plastic deformation ability, effectively absorbing the energy of crack propagation. This hinders continuous crack propagation and ultimately improves the toughness of the TiB_2-Ni composite film.

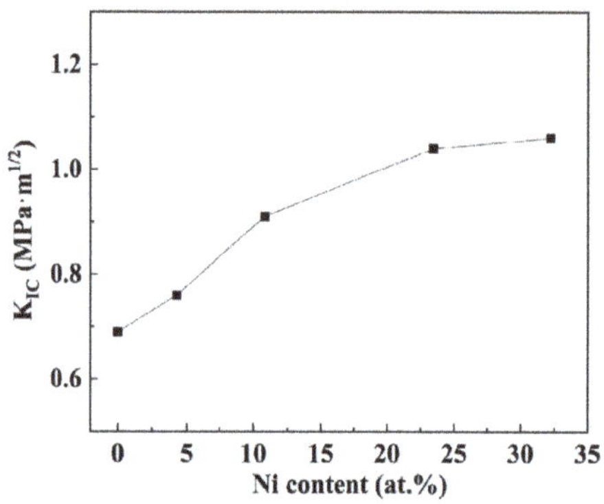

Figure 12. The calculated fracture toughness with different Ni contents.

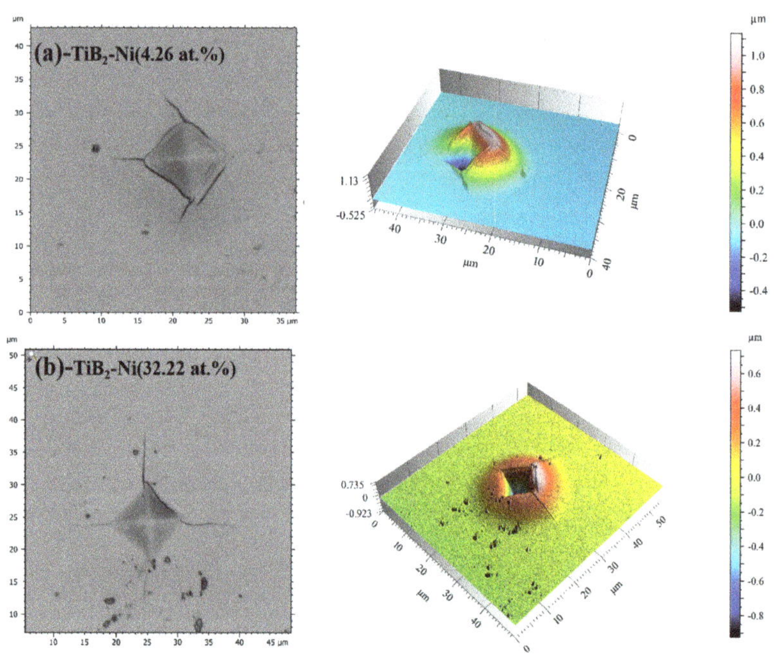

Figure 13. Indentation morphology of TiB_2 film with different Ni contents (**a**) 4.26 at.%;(**b**) 32.22 at.%.

Scratch testing was also used to evaluate the toughness of hard coatings by the references [25,47]. Figure 14 shows the scratch morphology with different Ni contents. As shown in the figure, the titanium diboride coating exhibits brittle characteristics. Shortly after testing, it develops cracks and subsequently spalls, which is indicative of the inherent brittleness associated with ceramic coatings. The addition of nickel delays the cracking time of the film. Owing to the accumulation of nickel in the amorphous phase at the boundary of titanium diboride, the soft metal can effectively absorb the energy associated with plastic deformation, and the microcracks initiated in TiB_2 are hindered by the surrounding nickel phase. The composite structure facilitates the passivation and deflection of cracks [48], thereby enhancing the toughness and binding force.

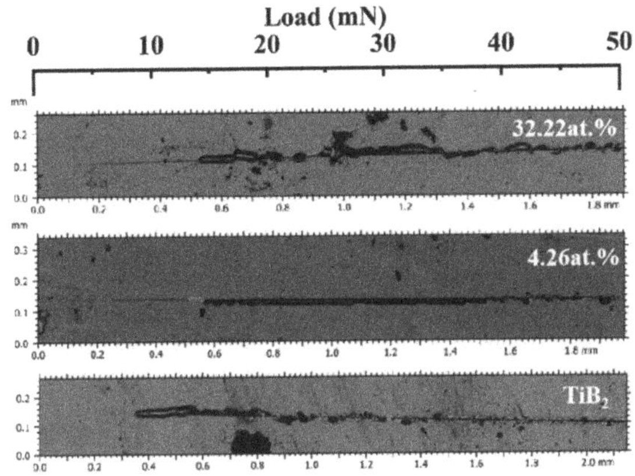

Figure 14. The nanoscratch morphology of TiB_2-Ni coatings.

Figure 15 illustrates the friction coefficient curve of the TiB_2-Ni composite coatings. For TiB_2 films, the friction coefficient remained consistently around 0.5. Upon the addition of nickel, the friction coefficient exhibited significant fluctuations during the initial stage but subsequently stabilized. Specifically, when the nickel content was 4.26 at.%, the friction coefficient stabilized at approximately 0.4. As the nickel content increased further, the friction coefficient demonstrated a gradual upward trend during the later stages of friction. The wear track morphology is shown in Figure 16. Titanium diboride coatings exhibit brittle spalling during wear, primarily due to the inherent brittleness of TiB_2 and insufficient adhesion between the coating and substrate. In contrast, a composite coating containing 4.26 at.% Ni demonstrated superior performance, exhibiting higher hardness, shallower wear marks, and reduced spalling compared to TiB_2 coatings. This improvement was attributed to the formation of a two-phase structure of TiB_2 particles surrounded by amorphous nickel. During plastic deformation under external forces, cracks were less likely to initiate at grain boundaries. The ductile nickel allowed it to absorb some of the deformation energy, effectively inhibiting crack initiation. Even if cracks did form, their propagation was hindered by mechanisms such as bypassing or cutting through the coating, further preventing crack extension. However, when the nickel content increased to 32.22 at.%, the coating's wear resistance significantly decreased, with evident plastic deformation observed in the wear tracks. This degradation in performance was likely due to the substantial reduction in hardness caused by the higher nickel content.

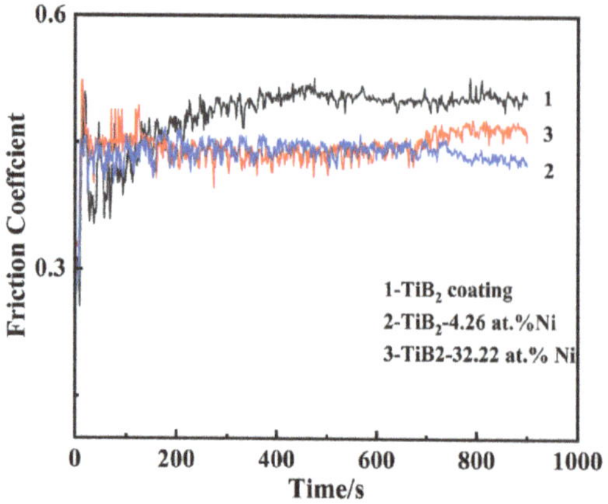

Figure 15. The friction coefficient of TiB_2-Ni composite coatings.

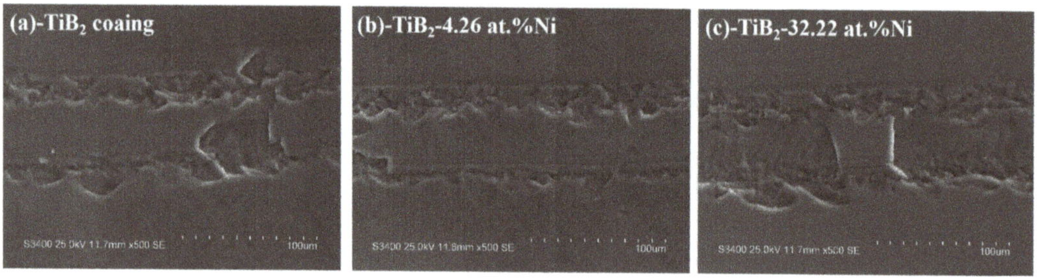

Figure 16. Wear track morphology of TiB_2-Ni coatings. (**a**) TiB_2 coating, (**b**)TiB_2-4.26 at.% Ni coating, (**c**) TiB_2-32.22 at.% Ni coating.

4. Conclusions

In this paper, TiB_2-Ni films were deposited using radio frequency magnetron sputtering. The study focused on the impact of Ni content on the composition, microstructure, and mechanical properties of the films. It was observed that as the radio frequency power of target Ni increased from 5 to 30 W, the nickel content in TiB_2 films also increased from 4.26 at.% to 32.22 at.%. The increase in Ni content led to grain refinement and a more compact and denser film structure. Diffraction peaks of (0001) for hexagonal TiB_2 were detected. With the addition of nickel content, the transition of titanium diboride to an amorphous state commenced, and no crystalline Ni or other Ni-containing phase was detected. Nickel existed at the boundary of titanium diboride as an amorphous phase, which formed a structure of titanium diboride encapsulated by the nickel layer. When the amount of nickel added is below 10.83 at%, the hardness of the coating can be maintained above 20 GPa. At a nickel content of 4.26 at.%, nanohardness was measured at approximately 34.8 $\pm$ 1.2 GPa, while elastic modulus was approximately 271 $\pm$ 15 GPa with high ratios of H/E and H^3/E^2. The enrichment of nickel in the grain boundary hindered the growth of titanium diboride and was beneficial to the enhancement of coating hardness. When the nickel content increased further, it became increasingly concentrated at the boundary, and titanium diboride tended to amorphous, which resulted in a decrease in the hardness of the film. The fracture toughness increased from 0.69 $MPa{\cdot}m^{1/2}$ to 1.06 $MPa{\cdot}m^{1/2}$ when the nickel content changed from 0 to 32.22 at.% in composite films. The addition of nickel enhanced the adhesion between the film and substrate while reducing the friction coefficient of the film. The Ni phase segregated in the boundary was found to prevent crack propagation through its superior plastic deformation ability, to achieve a toughening effect.

Author Contributions: Conceptualization, Y.W., H.S. and X.Q.; methodology, Y.W. and X.Q.; validation, X.W., X.L. and Y.Q.; formal analysis, Y.W. and X.Q.; investigation, Y.W. and X.Q.; writing—original draft preparation, Y.W.; writing—review and editing, X.Q. and N.Z.; visualization, Y.W. and X.Q.; supervision, H.S. and N.Z.; project administration, X.Q.; funding acquisition, Y.W. and X.Q. All authors have read and agreed to the published version of the manuscript.

Funding: This work was supported by the National Natural Science Foundation of China (No. 52002240 and 51401121), Shanghai Engineering Research Center of Hot Manufacturing (18DZ2253400) and Shanghai Collaborative Innovation Center of Manufacturing Technology for Heavy Casting and Forging.

Data Availability Statement: The authors confirm that the data supporting the findings of this study are available within the article.

Conflicts of Interest: The authors declare no conflicts of interest. The funders had no role in the design of the study; in the collection, analyses, or interpretation of data; in the writing of the manuscript; or in the decision to publish the results.

References

1. Holleck, H. Design of nanostructured thin films for tribological applications. In Proceedings of the TMS Conference Proceedings on Surface Engineering: Science and Technology I. Minerals, Metals, and Materials Society, Warrendale, PA, USA, 28 February–4 March 1999.
2. Cieniek, Ł.; Chudzik-Poliszak, E.; Moskalewicz, T.; Kopia, A.; Smolik, J. Effect of chromium doping on the substructure and mechanical properties of anti-wear TiB_2 coatings. *Arch. Civ. Mech. Eng.* **2023**, *23*, 80. [CrossRef]
3. Yang, T.; Wang, Z.; Tan, S.; Guo, F. Mechanical properties and growth mechanism of TiB_2-TiC/Fe composite coating fabricated in situ by laser cladding. *Appl. Compos. Mater.* **2020**, *27*, 877–893. [CrossRef]
4. Saeed, M.; Alexander, B.; Boyd, R.D.; Ali, D.; Schoemaecker, M.C.; Lionel, R. Impact of density on the behavior of suspension plasma sprayed TiB_2 coatings in the presence of molten aluminum. *J. Therm. Spray Technol.* **2022**, *31*, 1499–1507. [CrossRef]
5. Krasikov, A.V.; Markov, M.A.; Staritsyn, M.V.; Fedoseev, M.L.; Tkachev, D.A.; Bykova, A.D. Electrodeposition of Nickel-based composite electrochemical coatings using bulk-reinforced Al-TiB_2 powders. *Refract. Ind. Ceram.* **2023**, *64*, 444–448. [CrossRef]
6. Shimada, S.; Takahashi, M.; Kiyono, H.; Tsujino, J. Coatings and microstructures of monolithic TiB_2 films and double composite TiCN/TiB_2 films from alkoxide solutions by thermal plasma CVD. *Thin Solid Films* **2008**, *516*, 6616–6621. [CrossRef]
7. Fox-Rabinovich, G.S.; Gershman, I.S.; Veldhuis, S. Thin-film PVD coating metamaterials exhibiting similarities to natural processes under extreme tribological conditions. *Nanomaterials* **2020**, *10*, 1720. [CrossRef] [PubMed]
8. Duarte, A.; Coelho, B.; Vila, M.; Fernandes, A.J.S.; Silva, R.F.; Oliverira, F.; Costa, F.M. Deposition of TiB_2 onto X40 CrMoV 5-1-1 steel substrates by DC magnetron sputtering. *Vacuum* **2007**, *81*, 1519–1523. [CrossRef]
9. Bakhit, B.; Palisaitis, J.; Thörnberg, J.; Rosen, J.; Persson, P.O.Å.; Hultman, L.; Petrov, I.; Greene, J.E.; Greczynski, G. Improving the high-temperature oxidation resistance of TiB_2 thin films by alloying with Al. *Acta Mater.* **2020**, *196*, 677–689. [CrossRef]
10. Nedfors, N.; Primetzhofer, D.; Zhirkov, I.; Palisaitis, J.; Persson, P.O.Å.; Greene, J.E.; Petrov, I.; Rosen, J. The influence of pressure and magnetic field on the deposition of epitaxial TiBx thin films from DC magnetron sputtering. *Vacuum* **2020**, *177*, 109355. [CrossRef]
11. Mitterer, C. Borides in thin film technology. *J. Solid State Chem.* **1997**, *133*, 279–291. [CrossRef]
12. Hellgren, N.; Thörnberg, J.; Zhirkov, I.; Sortica, M.A.; Petrov, I.; Greene, J.E.; Hultman, L.; Rosen, J. High-power impulse magnetron sputter deposition of TiBx thin films: Effects of pressure and growth temperature. *Vacuum* **2019**, *169*, 108884. [CrossRef]
13. Palisaitis, J.; Dahlqvist, M.; Hall, A.J.; Thörnberg, J.; Persson, I.; Nedfors, N.; Hultman, L.; Greene, J.E.; Petrov, I.; Rosen, J.; et al. Where is the unpaired transition metal in substoichiometric diboride line compounds. *Acta Mater.* **2021**, *204*, 116510. [CrossRef]
14. Kunc, F.; Musil, J.; Mayrhofer, P.H.; Mitterer, C. Low-stress superhard Ti-B films prepared by magnetron sputtering. *Surf. Coat. Technol.* **2003**, *174–175*, 744–753. [CrossRef]
15. Berger, M.; Karlsson, L.; Larsson, M.; Hogmark, S. Low stress TiB_2 coatings with improved tribological properties. *Thin Solid Film.* **2001**, *401*, 179–186. [CrossRef]

16. Ding, J.C.; Zhang, T.F.; Yun, J.M.; Kim, K.H.; Wang, Q.M. Effect of Cu addition on the microsturcuture and properties of TiB_2 films deposited by a hybrid system combining high power impulse magnetron sputtering and pulsed dc magnetron sputtering. *Surf. Coat. Technol.* **2018**, *344*, 441–448. [CrossRef]
17. Sharma, R.; Singh, A.K.; Arora, A.; Pati, S.; De, P.S. Effect of friction stir processing on corrosion of Al-TiB_2 based composite in 3.5wt.% sodium chloride solution. *Trans. Nonferrous Met. Soc. China* **2019**, *29*, 1383–1392. [CrossRef]
18. Lin, W.S.; Chen, J.; Zhou, J. *Structure and Tribological Characterization of TiB_2/TiBN Multilayer Coatings Deposited by Magnetron Sputtering*; Springer: Berlin/Herdelberg, Germany, 2009. [CrossRef]
19. Lee, K.W.; Chung, Y.-W.; Korach, C.; Keer, L.M. Tribological and dry machining evaluation of superhard TiB_2/TiC multilayer coatings deposited on Si(001), M2 steel, and C3 WC cutting tool inserts using magnetron sputtering. *Surf. Coat. Technol.* **2005**, *194*, 184–189. [CrossRef]
20. Ma, B.; Sun, B.; Shang, H.; Li, R.; Cao, H.; Fernandes, F. Microstructure evolution and mechanical behavior of magnetron sputtering AlN-Al nanostructured composite film. *Ceram. Int.* **2024**, *50*, 6017–6024. [CrossRef]
21. Wolfe, D.E.; Singh, J.; Narasimhan, K. Synthesis and characterization of multilayered TiC/TiB_2 coatings deposited by ion beam assisted, electron beam-physical vapor deposition (EB-PVD). *Surf. Coat. Techology* **2003**, *165*, 8–25. [CrossRef]
22. Grigor'ev, O.N.; Koroteev, A.V.; Klimenko, A.V.; Mayboroda, E.E.; Prilutskii, É.V.; Bega, N.D. Fabrication and properties of multilayer ceramics in the SiC-TiB_2 system. *Refract. Ind. Ceram.* **2000**, *41*, 390–395. [CrossRef]
23. Stueber, M.; Holleck, H.; Leiste, H.; Seemann, K.; Ulrich, S.; Ziebert, C. Concepts for the design of advanced nanoscale PVD multilayer protective thin films. *J. Alloys Compd.* **2009**, *483*, 321–333. [CrossRef]
24. Abdollah, B.; Junko, U.; Ridvan, Y.; Hamidreza, G.; Issariyapat, A.; Bakar, T.A.A.; Kondoh, K. Deformation mechanism and enhanced properties of Cu-TiB_2 composites evaluated by the in-situ tensile test and microstructure characterization. *J. Alloys Compd.* **2020**, *847*, 156555. [CrossRef]
25. Wang, H.; Wang, B.; Li, S.; Xue, Q.; Huang, F. Tougheing magnetron sputtered TiB_2 coatings by Ni addition. *Surf. Coat. Technol.* **2013**, *232*, 767–774. [CrossRef]
26. Kang, Y.S.; Kang, S.H.; Kim, D.J. Effect of addition of Cr on the sintering of TiB_2 ceramics. *J. Mater. Sci.* **2005**, *40*, 4153–4155. [CrossRef]
27. Contreras, E.; Galindez, Y.; Gómez, M.A. Microstructure, mechanical and tribological properties of TiBC coatings by DC magnetron sputtering onto AISI M2 steel using independent TiB_2 and graphite targets. *Surf. Coat. Technol.* **2018**, *350*, 298–306. [CrossRef]
28. Chudzik-Poliszak, E.; Cieniek, Ł.; Moskalewicz, T.; Kowalski, K.; Kopia, A.; Smolik, J. Influence of W addition on microstructure and resistance to brittle cracking of TiB_2 coatings deposited by DCMS. *Materials* **2021**, *14*, 4664. [CrossRef] [PubMed]
29. Qiu, F.; Duan, X.; Dong, B.; Yang, H.; Lu, J.; Li, X. Effects of Cr and Zr addition on microstructures, compressive properties, and abrasive wear behaviors of in situ TiB_2/Cu cermets. *Materials* **2018**, *11*, 1464. [CrossRef] [PubMed]
30. Zhang, C.; Zhu, J.; Zhang, G.; Hu, Y. Laser powder bed fusion of nano-TiB_2 reinforced FeCoNiCr high-entropy alloy with enhanced strength and firm corrosion resistance. *J. Alloys Compd.* **2022**, *297*, 167110. [CrossRef]
31. Nedfors, N.; Mráz, S.; Palisaitis, J.; Persson, P.O.; Lind, H.; Kolozsvari, S.; Schneider, J.M.; Rosen, J. Influence of the Al concentration in Ti-Al-B coatings on microstructure and mechanical properties using combinatorial sputtering from a segmented TiB_2/AlB_2 target. *Surf. Coat. Technol.* **2019**, *364*, 89–98. [CrossRef]
32. Xian, L.; Li, L.; Fan, H.; Xian, G.; Zhao, H. Effect of doping Al, Ni, and Zr on the properties of TiB_2 coatings: A first-principle study. *Mater. Today Commun.* **2022**, *33*, 104844. [CrossRef]
33. Nakajima, H.; Koiwa, M.; Minonish, Y.; Ono, S. Diffusion of cobalt in single crystal α-Titanium. *Trans. Jpn. Inst. Met.* **1983**, *24*, 655–660. [CrossRef]
34. Mattias, B.; Coronel, E.; Olsson, E. Microstructure of d.c. magnetron sputtered TiB_2 coatings. *Surf. Coat. Technol.* **2004**, *185*, 240–244. [CrossRef]
35. Smolik, J.; Kacprzyńska-Golacka, J.; Sowa, S.; Piasek, A. The analysis of resistance to brittle cracking of tungsten doped TiB_2 coatings obtained by magnetron sputtering. *Coatings* **2020**, *10*, 807. [CrossRef]
36. Cai, X.; Xu, Y.; Zhong, L.; Liu, M. Fracture toughness of WC-Fe cermet in W-WC-Fe composite by nanoindentation. *J. Alloys Compd.* **2017**, *728*, 788–796. [CrossRef]
37. Hellgren, N.; Greczynski, G.; Sortica, M.A.; Petrov, I.; Hultman, L.; Rosen, J. X-ray photoelectron spectroscopy analysis of TiBx ($1.3 \leq x \leq 3.0$) thin films. *J. Vac. Sci. Technol. A* **2021**, *39*, 023403. [CrossRef]
38. Veprek, S.; Niederhofer, A.; Moto, K.; Bolom, T.; Männling, H.-D.; Nesladek, P.; Dollinger, G.; Bergmaier, A. Composition, nanostructure and origin of the ultrahardness in nc-TiN/a-Si_3N_4/a- and nc-$TiSi_2$ nanocomposites with Hv = 80 to $\geq$105GPa. *Surf. Coat. Technol.* **2000**, *133–134*, 152–159. [CrossRef]
39. Leyland, A.; Mattews, A. On the significance of the H/E ratio in wear control: A nanocomposite coating approach to optimised tribological behaviour. *Wear* **2000**, *246*, 1–11. [CrossRef]

40. Beake, B.D. The influence of the H/E ratio on wear resistance of coating systems- insights from small-scale testing. *Surf. Coat. Technol.* **2022**, *442*, 128272. [CrossRef]
41. Benzarti, Z.; Khalfallah, A.; Bougrioua, Z.; Evaristo, M.; Cavaleiro, A. Understanding the influence of physical properties on the mechanical characteristics of Mg-doped GaN thin films. *Mater. Chem. Phys.* **2023**, *307*, 128182. [CrossRef]
42. Musil, J. Hard nanocomposite coatings: Thermal stability, oxidation resistance and toughness. *Surf. Coat. Technol.* **2012**, *207*, 50–65. [CrossRef]
43. Zhang, S.; Yan, M.; Yang, Y.; Zhang, Y.; Yan, F.; Li, H. Excellent mechanical, tribological and anti-corrosive performance of novel Ti-DLC nanocomposite thin films prepared via magnetron sputtering method. *Carbon* **2019**, *151*, 136–147. [CrossRef]
44. Milman, Y.V.; Galanov, B.A.; Chugunova, S.I. Plasticity characteristic obtained through hardness measurement. *Acta Metall. Et Mater.* **1993**, *41*, 2523–2532. [CrossRef]
45. Boughrara, N.; Benzarti, Z.; Khalfallah, A.; Evaristo, M.; Cavaleiro, A. Comparative study on the nanomechanical behavior and physical properties influenced by the epitaxial growth mechanisms of GaN thin films. *Appl. Surf. Sci.* **2022**, *579*, 152188. [CrossRef]
46. Musil, J.; Jirout, M. Toughness of hard nanostructured ceramic thin films. *Surf. Coat. Technol.* **2007**, *201*, 5148–5152. [CrossRef]
47. Holmberg, K.; Laukkanen, A.; Ronkainen, H.; Wallin, K.; Varjus, S. A model for stresses, crack generation and fracture toughness calculation in scratched TiN-coated steel surfaces. *Wear* **2003**, *254*, 278–291. [CrossRef]
48. Pineau, A.; Benzerga, A.A.; Pardoen, T. Failure of metals I- brittle and ductile fracture. *Acta Mater.* **2016**, *107*, 424–483. [CrossRef]

nanomaterials

Article

Improved Ferroelectric and Magnetic Properties of Bismuth Ferrite-Based Ceramics by Introduction of Non-Isovalent Ions and Grain Engineering

Ting Wang [1,*], Huojuan Ye [1], Xiaoling Wang [1], Yuhan Cui [2], Haijuan Mei [1], Shenhua Song [3], Zhenting Zhao [1], Meng Wang [4], Pitcheri Rosaiah [5] and Qing Ma [3,*]

1 Guangdong Provincial Key Laboratory of Electronic Functional Materials and Devices, Huizhou University, Huizhou 516001, China
2 School of Mechanical and Electrical Engineering, Hebei Vocational College of Rail Transportation, Shijiazhuang 050801, China; cuiyhan@126.com
3 School of Materials Science and Engineering, Harbin Institute of Technology (Shenzhen), Shenzhen 518055, China
4 School of Undergraduate Education, Shenzhen Polytechnic University, Shenzhen 518055, China
5 Department of Physics, Saveetha School of Engineering, Saveetha Institute of Medical and Technical Sciences (SIMATS), Thandalam, Chennai 602105, India; rosaiah.p@gamil.com
* Correspondence: twang-hawk@foxmail.com (T.W.); wustmq@163.com (Q.M.)

Abstract: Single-phase multiferroics exhibiting ferroelectricity and ferromagnetism are considered pivotal for advancing next-generation multistate memories, spintronic devices, sensors, and logic devices. In this study, the magnetic and electric characteristics of bismuth ferrite ($BiFeO_3$) ceramics were enhanced through compositional design and grain engineering. $BiFeO_3$ ceramic was co-substituted by neodymium (Nd) and niobium (Nb), two non-isovalent elements, via the spark plasma sintering process using phase-pure powder prepared via sol-gel as the precursor. The symmetry of the sintered Nd–Nb co-doped samples changed from *R*3*c* to *Pnma*, accompanied by a decrease in the loss tangent, grain size, and leakage current density. The reduction in the leakage current density of the co-doped samples was ~three orders of magnitude. Moreover, ferroelectric, dielectric, and magnetic properties were substantially improved. The remanent polarization and magnetization values of the optimized Nd–Nb co-doped $BiFeO_3$ sample were 3.12 $\mu C\ cm^{-2}$ and 0.15 emu g^{-1}, respectively. The multiferroic properties were enhanced based on multiple factors such as structural distortion caused by co-doping, grain size reduction, suppression of defect charges via donor doping, space-modulated spin structure disruption, and an increase in magnetic ions. The synergistic approach of composition design and grain engineering sets a paradigm for the advancement of multiferroic materials.

Keywords: bismuth ferrite; spark plasma sintering; Nd–Nb co-doping; ferroelectric property; magnetic property

1. Introduction

Compounds with $\geq$2 primary ferroic orders (such as ferromagnetism, ferroelasticity, and ferroelectricity) exhibit multiferroic characteristics. These compounds have been technologically and economically important over the past two decades, and continue to be now and will be in the future [1]. The effective magnetoelectric coupling at room temperature has several applications in data storage systems, adjustable microwave components, sensors, and random-access memory devices [2–4]. Magnetoelectric multiferroic materials

 https://doi.org/10.3390/nano15030215

are of three types: (i) single-phase type I multiferroics with distinct magnetization and polarization origin, (ii) single-phase type II multiferroics with polarization induced by a particular magnetic configuration, and (iii) composite materials or heterostructures (piezoelectric multilayer and artificial ferromagnetic capacitors) with strain-mediated interactions modulating magnetic properties under an electric field (E) [1].

Bismuth ferrite ($BiFeO_3$, BF) has emerged as the quintessential example of single-phase type I multiferroic materials at room temperature since 2003 owing to its substantial ferroelectric polarization and elevated phase transition temperatures (Curie temperature (T_C) of ~825 °C and Néel temperature (T_N) of ~370 °C) [5]. It exhibits a distorted rhombohedral perovskite structure belonging to the $R3c$ space group with lattice parameter values of length (a) and angle (α) of 3.958 Å and 89.30°, respectively [6]. It is characterized by better ferroelectric properties than lead-free materials such as bismuth titanate ($Bi_4Ti_3O_{12}$) [7], with simultaneous antiferromagnetic and ferroelectric characteristics at ambient temperature [6]. BF crystals exhibit ferromagnetism and ferroelectricity owing to the partially filled 3D orbits of ferric ions (Fe^{3+}) and the stereochemically active lone-pair electrons in the $6s^2$ orbitals of bismuth ions (Bi^{3+}) [8]. The theoretical ferroelectric polarization value of BF in the [111] direction is in the range of 90–100 μC cm^{-2} [9], while the magnetoelectric effect is prominent with the coupling energy of ~1.25 meV [10]. Although exceptional electrical properties of BF are predicted theoretically, it is practically challenging owing to the high leakage current density (J). This is attributed to oxygen vacancy, Bi evaporation, the formation of the frequent secondary phase, the presence of mixed Fe^{3+} and ferrous (Fe^{2+}) ions, and weak ferromagnetism, which substantially limit its application in various sophisticated devices [10]. The central challenge for $BiFeO_3$-based multiferroic ceramics has been to concurrently enhance ferroelectricity and ferromagnetism, thereby improving the magnetoelectric coupling.

Chemical substitution at the A-, B-, or A/B-sites of BF is an effective method to increase its ferroelectric properties by reducing leakage current [1–4,11]. Lakshmi et al. [12] reported a substantial reduction in the leakage current of BF ceramics doped with niobium (Nb) of high valency. The doping of Nb could alter the Fe-O-Fe bond angle, potentially disrupting the spin cycloid structure. Abe et al. [13] examined the effects of cation substitution with different valences in $BiFeO_3$ ceramics and observed a marked reduction in leakage current by substituting BF with 10% titanium ion (Ti^{4+}), which was attributed to a decrease in oxygen vacancies. Yamamoto et al. [14] identified a transition from a cycloidal spin structure to a collinear spin structure of BF polycrystalline ceramics substituted by cobalt (Co), which induced spontaneous magnetization at room temperature owing to the spin-canting phenomenon occurring in collinear structures. The doping of A-sites with rare-earth ions is a beneficial strategy. Studies indicated that the crystal structure was stabilized via the partial replacement of the Bi^{3+} ions of BF with rare-earth ions by enhancing ionic bond strength, which suppressed secondary phases and oxygen vacancies primarily formed by the evaporation of Bi. Moreover, magnetocrystalline anisotropy increased, which was responsible for the energetically unfavorable spin cycloid structure [15–20]. Recently, the symmetry modulation of $R3c$ to $Pna2_1$ and eventually to $Pbnm$ with the increasing substitution of lanthanum (La)- [19], neodymium (Nd)- [16], samarium (Sm)- [20], gadolinium (Gd)- [18], and La/lutetium (Lu) [17] was developed as an effective method to improve the multiferroic characteristics of BF ceramics. The emergence of the $Pna2_1$ polar phase under chemical pressure is a determining factor in this process. Quan et al. [15] reported the improved magnetic properties of $Bi_{0.8}Ce_{0.2}FeO_3$, which were attributed to the high canting angle and repressed cycloidal spin structure of the cerium (Ce)-doped sample. Nonzero magnetization was observed in La-substituted BF [21], a phenomenon owing to the chemical compression-induced rotation of iron oxide (FeO_6) octahedra. In the FeO_6 octahedra, each Fe^{3+} is surrounded by six O^{2-} and the two nearest neighboring Fe^{3+} are connected by an O^{2-}, where the Fe-3d energy level and O-2p energy level hybridizing contributes to double-exchange interaction. The Fe-O-Fe bond angle changed with the doping La element, which was helpful for the exchange interaction by enhancing the overlap of the Fe 3d-O 2p orbitals. Thus, these modified the symmetry structure and suppressed the spin cycloid. Thus, the presence of rare-earth ions alters chemical pressure and disrupts the spin cycloid, progressively enhancing magnetic properties.

Hence, it was anticipated that co-doping the Fe and Bi sites of BF might decrease J while simultaneously enhancing magnetic and ferroelectric properties. Ke et al. [22] co-doped BF ceramics using Nd and titanium (Ti) to obtain structures with morphotropic phase boundaries (MPB) using a solid-state sintering method. The synthesized samples exhibited improved magnetic and ferroelectric properties at the boundaries with the highest remanent magnetization (M_r) and polarization (P_r) values of 0.27 emu g^{-1} and 29 $\mu C\ cm^{-2}$, respectively. Gumiel et al. [23] reported that antiferromagnetism could be converted to weak ferromagnetism via co-doping (Nd-Ti), which also decreased electric conductivity, increased the electric polarization, and stabilized the converted state. Wang et al. [24] co-doped BF ceramics with Sm and Nb via spark plasma sintering and the sol-gel process. The doped samples exhibited reduced J with enhanced magnetic and ferroelectric properties. Gao et al. [25] reported the synthesis of polar nanoregions by co-doping BF with Sm and scandium (Sc), which increased the dielectric breakdown field. Nonetheless, reports on the improved magnetic and ferroelectric properties and the leakage current in $BiFeO_3$ ceramics co-doped with Nd and high-valence Nb are scarce.

On the other hand, the technique used to fabricate ceramics significantly influences their macroscopic multiferroic characteristics. Several methods, such as rapid liquid-phase sintering [26], the sol-gel method [27], electric current activated sintering [28], flash sintering [29], and microwave sintering [30] were used to prepare single-phase BF. The spark plasma sintering (SPS) method is considered a more efficient technique than the aforementioned techniques for synthesizing materials with micro-/nanostructures that can be modulated. The pulsed electric current and uniaxial stress generated during SPS facilitate rapid sintering and effective heat transfer, inducing enhanced ceramic compaction within a short duration [31]. It is characterized by low sintering temperatures and less holding time, which inhibit Bi volatilization and the valence state change of Fe. SPS was used to successfully fabricate highly dense BF and BF-based ceramics with a pure phase and enhanced multiferroic and dielectric properties [24,31,32]. In this context, the synergistic impact of co-substitution in ferroelectric ceramics can be beneficial for enhancing physical properties. This inspired us to use SPS to prepare nanocrystalline $BiFeO_3$ ceramics co-doped with rare-earth Nd^{3+} and high-valence Nb^{5+}.

In this study, inspired by the concept of the grain size and ionic radius effect, the proposed strategy involves the co-substitution of non-isovalent ions (Nd and Nb) at the A/B sites via the sol-gel method followed by SPS to enhance the ferroelectric and magnetic characteristics of BF ceramics. The $Bi_{1-x}Nd_xFe_{0.95}Nb_{0.05}O_3$ (x = 0, 0.05, 0.10, 0.15, and 0.20) ceramic samples were sintered using phase-pure sol-gel-derived precursor powders. The impact of Nd–Nb co-doping on the ferroelectric, structural, dielectric, and magnetic characteristics of the synthesized samples was thoroughly investigated.

2. Experimental Procedure

BF and $Bi_{1-x}Nd_xFe_{0.95}Nb_{0.05}O_3$ (x = 0.05, 0.10, 0.15, and 0.20) nanopowder samples in a pure phase were synthesized using the sol-gel method, as documented in our earlier studies [33,34]. These as-synthesized powder samples were loaded into a 10 mm-diameter graphite mode and sintered via SPS. SPS was performed following the reported procedure [24,35]. The pellet was sintered at 700 °C, with a heating and cooling rate of 100 °C/min. The temperature was held there for 5 min under a vacuum of 10^{-2} Pa. A uniaxial and stable stress of 50 MPa was maintained both in the heating and soaking processes. The $Bi_{1-x}Nd_xFe_{0.95}Nb_{0.05}O_3$ samples with x values of 0.05, 0.10, 0.15, and 0.20 were denoted as 5NdNb, 10NdNb, 15NdNb, and 20NdNb, respectively.

The phase composition of the sintered ceramic samples via SPS was examined via X-ray diffraction (XRD). The microstructures of the fractured surfaces of the ceramic samples were analyzed using a field emission–scanning electron microscope (FE-SEM, model Hitachi S-4700). The samples were ground to obtain 1 mm thickness, coated with silver electrodes, and subsequently annealed for 2 h at 120 °C to analyze the dielectric properties using a computerized impedance analyzer (PSM1735, Newton 4th Ltd., UK). Polarization versus electric field (*P–E*) and *J* were measured using a ferroelectric testing system (Premier-II by Radiant Technologies, Inc.). Magnetic properties were analyzed using a physical property measurement system (PPMS, DynaCool-9T, by QUANTUM DESIGN). All experimental data were recorded at ambient temperature.

3. Results and Discussion

Figure 1a presents the XRD patterns of pristine and Nd–Nb co-substituted $BiFeO_3$ ceramic samples. Phase-pure BF ceramics were effectively synthesized via SPS in conjunction with the powder samples obtained via the sol-gel method [24]. The diffraction peaks of (104), (110), and (113) were indexed to a characteristic perovskite structure distorted rhombohedrally with the *R3c* space group [24]. Impure peaks in the doped ceramics were absent, and the sharpness of the diffraction peaks indicates high crystallinity. No additional peaks were observed for ceramics up to a co-doped level of 0.10, suggesting *R3c* as the primary phase structure. The multiple peaks at 2θ of ~32°, 40°, 52°, and 57° merged to form a broad peak, indicating the successful doping of Nd and Nb in BF [36]. The magnified patterns at 2θ of ~32° presented in Figure 1b show the overlapping of the split peaks (104) and (110). Moreover, the peaks shifted to high diffraction angles, indicating a reduction in the unit cell size and lattice constant owing to the smaller ionic radius of Nd^{3+} than that of Bi^{3+} and the similar ionic radii of Nb^{5+} and Fe^{3+} (Nb^{5+} = 0.64 Å, Nd^{3+} = 0.983 Å, Fe^{3+} = 0.645 Å, and Bi^{3+} = 1.173 Å) [34,37]. This observation further confirms the successful integration of Nb and Nd within BF. A few peaks of 15NdNb and 20NdNb re-split to form new peaks, particularly for 20NdNb, suggesting the presence of a mixture of orthorhombic *Pnma* and rhombohedral *R3c* phases, conforming to the reported results [37]. Therefore, it can be inferred that Bi and Fe substituted by Nd and Nb, respectively, substantially altered the crystal structure and lattice parameters of BF.

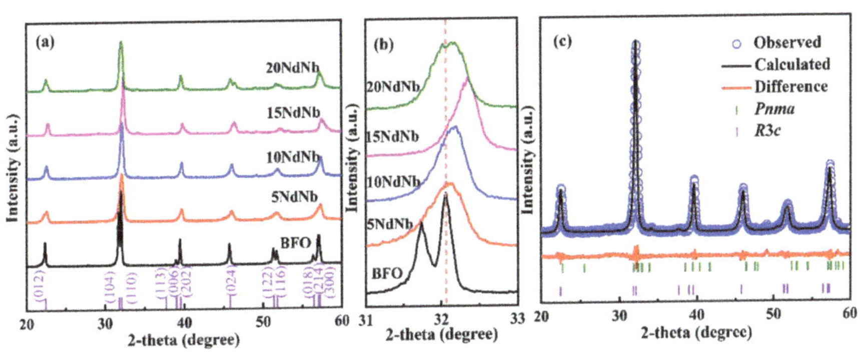

Figure 1. (**a**) XRD patterns of undoped and Nd–Nb co-doped $BiFeO_3$ ceramics; (**b**) magnified patterns around 2θ ~32°; (**c**) Rietveld refinement XRD patterns of 10NdNb samples.

The quantitative correlation between structural transition and doping concentration was elucidated via Rietveld refinement using the General Structure Analysis System (GSAS) program coupled with the graphical user interface (EXPGUI). Figure 1c presents the Rietveld refinement of the XRD patterns of 10NdNb. The refined lattice parameters, phase constitution, and *R*-factors of the samples are listed in Table 1. The minimal discrepancy between the observed and calculated data, along with GOF (goodness of fit, R_{wp}/R_{exp}) values ranging from 1.3 to 1.6, confirms the success of the refinement process. The determined *a* and *c* of the pristine BF ceramics were 5.571 and 13.872 Å, respectively, which align with the *R3c* space group (JCPDS No. 86-1518) [24]. The refinement considered two phases for Nd–Nb co-doped BF, such as *R3c* + *Pbnm*, *R3c* + *Pbam*, or *R3c* + *Pnma*. The *R3c* + *Pnma* model yielded the most accurate fit. The *R3c* fraction decreased from 97.45% to 75.12% upon Nd–Nb co-doping. Lattice parameters *a* and *c* exhibited a steady increase and decrease, respectively, which can be ascribed to the different ionic radii of Fe, Bi, Nd, and Nb and the change in bonding angle. This explains the observed shift in diffraction peaks to large angles in Figure 1b (for samples: BFO, 5NdNb, 10NdNb, and 15NdNb).

Table 1. Phase constitution, lattice parameters, and *R*-factors estimated by Rietveld refinement for undoped and Nd–Nb co-doped samples.

Samples	Space Group	Phase Constitution (%)	Lattice Parameters (Å)	*R*-Factors
BFO	*R3c*	100	$a = 5.571$ $c = 13.872$	$R_{exp} = 7.36$ $R_{wp} = 11.34$ GOF = 1.54
5NdNb	*R3c*	97.45	$a = 5.578$ $c = 13.821$	$R_{exp} = 6.57$ $R_{wp} = 8.67$ GOF = 1.32
	Pnma	2.55	$a = 5.623$ $b = 7.946$ $c = 5.858$	
10NdNb	*R3c*	90.29	$a = 5.587$ $c = 13.796$	$R_{exp} = 5.82$ $R_{wp} = 7.68$ GOF = 1.32
	Pnma	9.71	$a = 5.572$ $b = 7.874$ $c = 5.869$	
15NdNb	*R3c*	85.46	$a = 5.592$ $c = 13.774$	$R_{exp} = 6.13$ $R_{wp} = 8.65$ GOF = 1.41
	Pnma	14.54	$a = 5.569$ $b = 7.866$ $c = 5.883$	
20NdNb	*R3c*	75.12	$a = 5.597$ $c = 13.721$	$R_{exp} = 5.59$ $R_{wp} = 7.95$ GOF = 1.42
	Pnma	24.88	$a = 5.546$ $b = 7.810$ $c = 5.912$	

To examine the surface morphology and grain size variations in the ceramics, SEM was utilized. Figure 2 illustrates the distinct grain structure of Nd–Nb co-doped BF samples. Analytical software (Nano Measurer 1.2, Department of Chemistry, Fudan University, Fudan, China) was employed to calculate the grain-size distributions of the as-prepared ceramics. We reported earlier that undoped BFO comprises large polyhedral grains with sizes of ~1–3 μm [24]. The obtained average grain size of BFO samples was about 2.16 μm (Figure 2a). A few pores existed at grain boundaries owing to oxygen vacancies and Bi volatilization [38]. In contrast, the Nd–Nb co-doped samples exhibited flatter boundary facets with no evident pores and a uniform distribution. The gradual reduction in grain size with increasing Nd–Nb co-doping is evident in Figure 2b–e. The grain size diminished with the increase in the co-doped content, reaching an average size <300 nm of 20NdNb, indicating that Nd–Nb co-doping effectively inhibits grain growth. It is known that the presence of vacancies in oxide materials is beneficial for ion transport during sintering and thus generates larger grains for ceramic. The marked decrease in grain size can be attributed to the low diffusivity of Nd^{3+} and the suppression of Bi volatilization and oxygen vacancy formation because of Nd–Nb co-doping [39]. Figure 2g,h illustrates the surface scan results, showing even distribution of elements within the scanned area without non-uniform aggregation or segregation. Generally, the microstructure significantly impacts the macroscopic properties, as documented in studies on ferroelectric oxides [1,2,40]. The influence of grain size on magnetic properties in $BiFeO_3$ ceramics has been noted in single-phase $BiFeO_3$ [41]. Consequently, with the pronounced changes in surface morphology induced by Nd–Nb co-doping, the multiferroic characteristics and electronic structure of Nd–Nb co-doped BF differ from those of the pristine BF, which is discussed in the subsequent section.

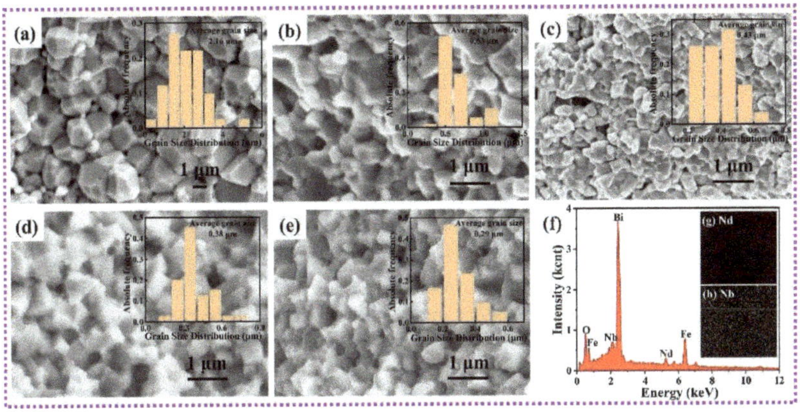

Figure 2. SEM micrographs of the fresh fracture surfaces for (**a**) BFO, (**b**) 5NdNb, (**c**) 10NdNb, (**d**) 15NdNb, (**e**) 20NdNb, and (**f**) EDS; (**g**) Nd element mapping; and (**h**) Nb element mapping for 15NdNb samples (inset: the average grain size).

Figure 3 depicts the change in dielectric behavior (dielectric constant and dielectric loss) of pristine BF and Nd–Nb co-doped BF with temperature at 1 MHz frequency. The dielectric constant was independent of temperature below 200 °C, signifying the stability of the ferroelectric order. This stability is attributed to the high ferroelectric T_C and the unique micro- and electronic structures of BF [42]. The effect of lattice vibrations on electron movement diminished at high frequencies, leading to a less pronounced variation in dielectric properties with temperature at these frequencies. The dielectric constant increased gradually with a further increase in temperature owing to thermally activated ionic conductivity caused by an increase in defects, such as oxygen vacancies, under higher temperatures [16,19]. The loss tangent of BF increased with increasing temperature, whereas the same for Nd–Nb co-doped BF remained relatively stable below 200 °C. The dielectric constant of Nd–Nb co-doped BF increased, while the loss tangent exhibited an opposite trend to those of BF, suggesting that the dielectric characteristics of Nd–Nb co-doped BF were altered. The dielectric constants of BF, 5NdNb, 10NdNb, 15NdNb, and 20NdNb at room temperature were 143, 176, 187, 211, and 132, respectively. The SEM images of Nd–Nb co-doped BF indicate that the grain sizes of Nd–Nb co-doped samples were reduced, which increased the grain boundary. Additionally, Nd–Nb co-doping could induce phase transition and hinder Bi volatilization. Thus, resistivity was increased and the oxygen vacancy concentration was decreased [26]. Consequently, the dielectric constant was enhanced, while the loss tangent was reduced. Although the loss tangent remained slightly high, the incorporation of other ABO_3 materials and manganese oxide (MnO_2), or by subjecting the samples to annealing under various atmospheric conditions, could potentially have reduced the dielectric loss [43].

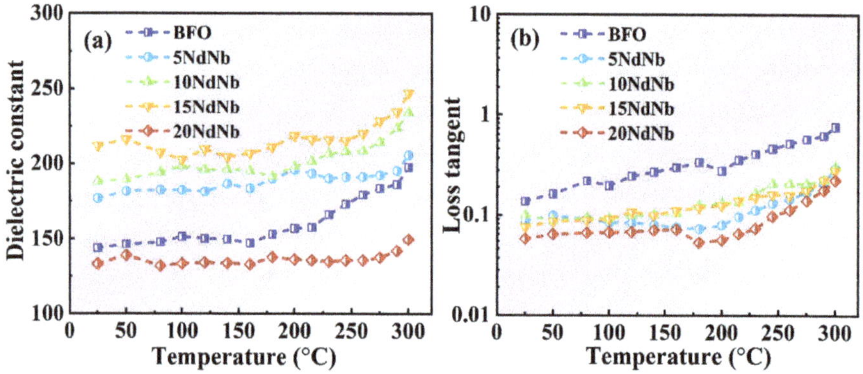

Figure 3. Temperature dependences of (**a**) dielectric constant and (**b**) loss tangent at 1 MHz for undoped and Nd–Nb co-doped $BiFeO_3$ ceramics.

The ferroelectric properties of the synthesized ceramics were investigated by measuring ferroelectric hysteresis (*P*–*E*) loops at room temperature and 10 Hz using a ferroelectric analyzer [Figure 4a]. The P_r and coercivity field (E_c) values of the ceramics are presented in Figure 4b and Table 2. The results were directly compared based on the magnitude, as the measurements were conducted using consistent parameters. Figure 4a shows no inflation of polarization loops, although they were not completely saturated, indicating the production of high-quality ceramics and a marked reduction in the leaky behavior of BFO. BF ceramics exhibited unsaturated and substantially inflated *P*–*E* loop area owing to a high *J* [31]. Furthermore, it is noteworthy that the ferroelectric polarization was highly dependent on the Nd–Nb concentration, with a substantial enhancement observed for the 10NdNb ceramics (see Figure 4a). Notably, E_c and P_r exhibited nonlinear behavior, as illustrated in Figure 4b. BF exhibited P_r and E_c values of ~0.41 μC cm^{-2} and ~4.45 kV cm^{-1}, respectively. Nd–Nb doping substantially improved the shape of *P*–*E* loops. The P_r changed with the doping content, exhibiting an optimal value of ~3.12 μC cm^{-2} for 10NdNb, which is considerably higher than that of BF (0.41 μC cm^{-2}). Thus, Nd–Nb co-doping is an appropriate strategy for enhancing the ferroelectric characteristics of BF ceramics. The substitution of Bi and Fe sites of BF by Nd^{3+} and Nb^{5+} ions, respectively, altered the grain size and structural characteristics of the crystals, affecting the E_c and P_r of the materials.

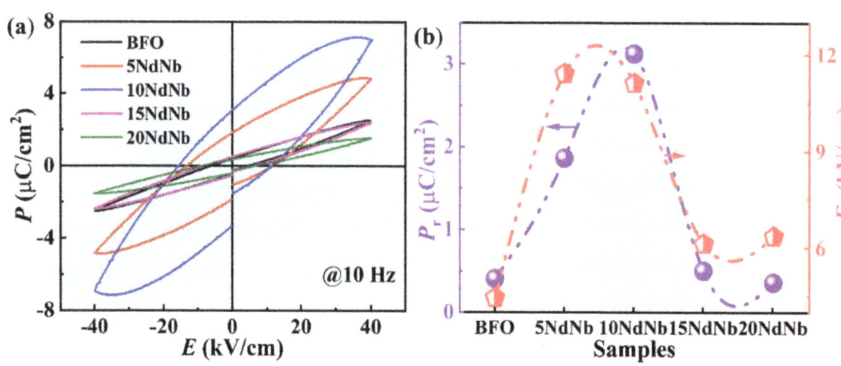

Figure 4. (**a**) Room−temperature ferroelectric hysteresis loops (*P*−*E*) and (**b**) ferroelectric parameters determined at 10 Hz for undoped and Nd–Nb co-doped $BiFeO_3$ ceramics (P_r: remnant polarization and E_c: coercive field).

Table 2. Ferroelectric and magnetic parameters for undoped and Nd–Nb co-doped $BiFeO_3$ ceramics.

Parameter	BFO	5NdNb	10NdNb	15NdNb	20NdNb
P_r ($\mu C/cm^2$)	0.41	1.86	3.12	0.51	0.36
E_c (kV/cm)	4.45	11.40	11.10	6.13	6.36
M_r (emu/g)	5.67×10^{-4}	0.070	0.12	0.15	0.14
H_c (kOe)	0.050	6.17	8.84	9.80	9.05

In the doped samples, some Bi sites were replaced by Nd, and the lone-pair electrons of Bi^{3+} ions (s^2 electrons) hybridized with the empty p orbitals of Bi^{3+} or O^{2-} ions, forming a localized lobe. This phenomenon induces non-centrosymmetric distortion, thereby enhancing ferroelectricity. Nd–Nb co-doping is also responsible for changing lattice parameters. Consequently, the improved ferroelectricity was attributed to the decreased concentration of oxygen vacancy, and even the decrease in the amount of Fe^{2+}, which helps to mitigate domain-pinning effects [44]. Additionally, the grain boundaries in the rhombic-phase ceramics impeded the movement of electric domains and polarization reversal, thereby enhancing ferroelectric polarization [42]. However, samples doped with highly concentrated dopants ($x > 0.10$) increased the *Pnma* phase. 10NdNb represents the optimal sample based on its position at the morphotropic-phase boundary [45]. The reduced ferroelectricity of 15NdNb might be attributed to several factors: (i) the stereochemically active $6s^2$ lone-pair electrons of Bi^{3+} and empty 6p orbits induce the ferroelectricity of BF, so when the doping content exceeds a certain value, Nd-doped Bi sites decrease the concentration of lone-pair electrons (s^2 electrons) and deteriorate the stereochemical activity of the lone pair owing to the spherically distributed electron density of Nd^{3+} [36]; (ii) the highly centrosymmetric orthorhombic phase fails to exhibit ferroelectricity [31] and induce polarization easily; or (iii) a high Nd concentration indicates a low ability of the cations to displace within the crystal lattice, resulting in a low ion displacement and ferroelectric polarization [44]. Wu et al. [42] investigated Tb-doped $BiFeO_3$ ($x = 0.05$ and 0.1) synthesized via an optimized sol-gel method and reported a maximum P_r of 6.9 $\mu C/cm^2$. Islam et al. [38] prepared $Bi_{0.8}Ba_{0.2}Fe_{0.9}Ta_{0.1}O_3$ ceramic through a solid-state reaction, achieving a P_r of approximately 0.15 $\mu C/cm^2$. Zhang et al. [46] reported a P_r of 0.75 $\mu C/cm^2$ for $Bi_{0.95}Dy_{0.05}Fe_{0.95}Mn_{0.05}O_3$ ceramics prepared by a solid-state reaction. Hua et al. [47] reported that the P_r of $Bi_{0.925}Ho_{0.075}Fe_{0.95}Mn_{0.05}O_3$ ceramic was measured to be 0.0948 $\mu C/cm^2$. Our values are comparably favorable.

Figure 5a illustrates the J versus E characteristics of the as-sintered ceramics measured at ambient temperature. The plots exhibit symmetry under negative and positive biases. The J of BF, 5NdNb, 10NdNb, 15NdNb, and 20NdNb ceramics at an electric field of ± 30 kV cm^{-1} were ~1.26×10^{-5}, 3.39×10^{-7}, 1.78×10^{-7}, 5.82×10^{-8}, and 9.51×10^{-8} A cm^{-2}, respectively. Thus, Nd–Nb co-doped BF ceramics were remarkably lower J than BF under identical electric fields. The smallest J of Nd–Nb co-doped BF was ~three orders of magnitude lower than that of BF, suggesting superior insulating properties. This improvement can be ascribed to the reduced Bi volatility, charge defects (oxygen or Bi vacancies), and Fe^{2+} content [2]. The bond enthalpy values of Nb–O and Nd–O were higher than those of Fe–O and Bi–O, stabilizing the perovskite structure through the substitution of Bi and Fe by Nd and Nb, respectively, which also diminished Bi evaporation [39,48]. Moreover, the substitution of Nb^{5+} ions, which had a higher valence, for Fe^{3+} ions, helped to occupy oxygen vacancies. Consequently, the co-doping of Nd and Nb effectively diminished the leakage current density in these ceramic samples.

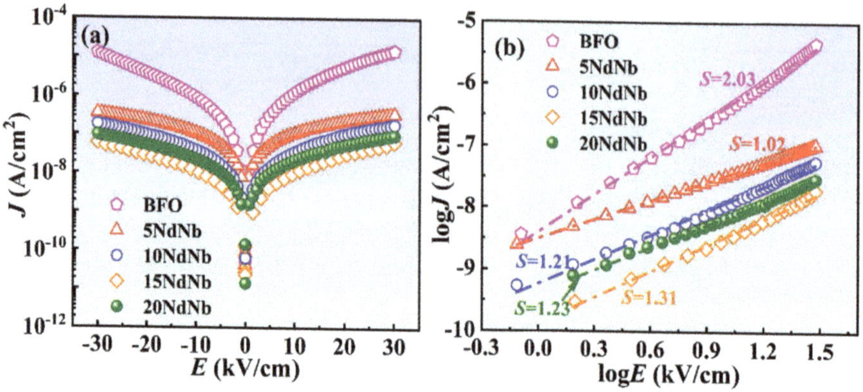

Figure 5. (**a**) Leakage current density as a function of electric field ($J-E$) and (**b**) leakage mechanism for undoped and Nd–Nb co-doped $BiFeO_3$ ceramics, determined at room temperature.

Five primary leakage mechanisms of perovskite ferroelectric materials have been identified: Schottky emission, space–charge limited current (SCLC), Poole–Frenkel emission, ohmic conduction, and Fowler–Nordheim (FN) tunneling [35]. Specifically, ohmic conduction and SCLC are the predominant leakage mechanisms of BF-based materials. Leakage current behavior can be modeled by applying the equation $J_{SCLC} \propto E^n$. The leakage behavior follows SCLC and ohmic conduction mechanisms when exponent (n) values are 2 and 1, respectively. The leakage data were reanalyzed by plotting log J against log E, as depicted in Figure 5b. The value of n of BF was ~2, suggesting that leakage was predominantly owing to the SCLC mechanism. The slope (n) of Nd–Nb co-doped BF lay between 1 and 2, indicating combined SCLC and ohmic conduction mechanisms. Thus, the slope of BF decreased with the introduction of Nd–Nb, which implies a reduction in mobile charge carriers due to enhanced structural stability [44].

Figure 6 displays the magnetic hysteresis loops of the prepared ceramic samples at ambient temperature. The M_r and coercivity (H_c) data of each sample are presented in Figure 6b and Table 2. None of the loops reached saturation under the magnetic field of 50 kOe. The magnetic moment of BF was nearly negligible owing to its G-type antiferromagnetic character, similar to reported studies [49,50]. BF exhibited a nearly linear loop with a negligible M_r (~5.67×10^{-4} emu g^{-1}), characteristic of antiferromagnetic materials, which is in agreement with other findings [49,50]. The loop centers of Nd–Nb co-doped BF expanded, indicating a shift towards weak ferromagnetism and a substantial increase in M_r, suggesting a transition from an antiferromagnetic state to a ferromagnetic state [24]. The M_r values of 5NdNb, 10NdNb, 15NdNb, and 20NdNb were ~0.07, 0.12, 0.15, and 0.14 emu g^{-1}, respectively, which are over two orders of magnitude higher than those of BF. This enhancement is attributed to structural distortion. Nd–Nb co-doped BF ceramics induce a structural transformation with the coexistence of rhombohedral and orthorhombic phases. The equilibrium between antiparallel spin lattices of Fe^{3+} adjacent to each other is disrupted because of the partially substituted Bi^{3+} and Fe^{3+} by Nd^{3+} and Nb^{5+}, respectively, with different ionic radii and valencies. This symmetry distortion typically disrupts the spatially non-uniform spin helical structure, altering bond lengths and angles, which in turn increases the spin tilt angle and results in a net macroscopic magnetization [15]. The symmetry-driven Dzyaloshinsky–Moriya (DM) interaction mechanism suggests that the tilted antiferromagnetic spin orders demonstrate enhanced magnetic characteristics [16]. Concurrently, the unique spin–helix long-range ordered structure in $BiFeO_3$ is disrupted, liberating additional magnetic moments [51], which couples with increased polarization. The high coercivity is linked to the magnetic anisotropy induced by co-doping. As documented by Monem and colleagues [52], the peak M_r for $Bi_{0.93}Sr_{0.07}Fe_{0.8}Zr_{0.2}O_3$ ceramic, synthesized via a tartaric acid-assisted sol-gel method, reaches approximately 3.67 emu/g. Yadav et al. [53] and Rao et al. [39] separately synthesized $Bi_{0.9}Nd_{0.1}Fe_{1-x}Ti_xO_3$ and $Bi_{0.85}Sm_{0.15}Fe_{0.9}Sc_{0.1}O_3$ ceramics through conventional solid-state reactions, obtaining M_r values of about 3.15 and 0.2 emu/g, respectively. Evidently, the current findings are in good agreement with these studies. Without a doubt, the current results suggest that the approach of compositional design coupled with grain engineering is a promising strategy for enhancing the multiferroic properties of $BiFeO_3$ at room temperature.

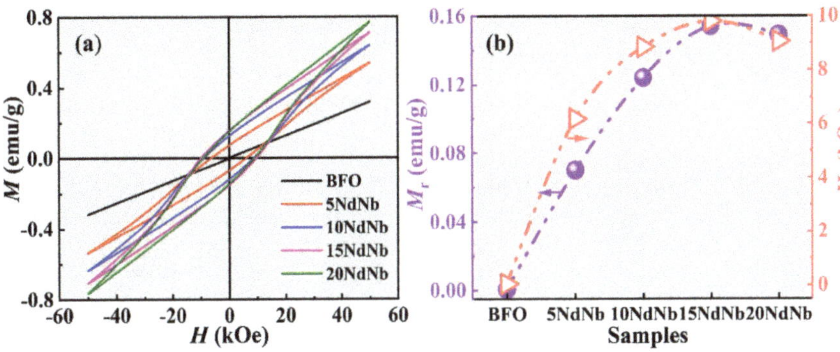

Figure 6. (**a**) Magnetic hysteresis loops ($M-H$) and (**b**) ferromagnetic parameters (remnant magnetization (M_r) and coercive field (H_c)) for undoped and Nd–Nb co-doped $BiFeO_3$ ceramics, determined at room temperature.

4. Conclusions

Phase-pure multiferroic Nd–Nb co-doped BF ceramics were successfully synthesized via spark plasma sintering. The phase composition of BF ceramics was altered from an exclusive *R3c* phase to a mixed phase comprising *R3c* and *Pbnm* phases via Nd–Nb co-doping. The reduced-loss tangent and J were correlated to the decrease in the defect concentration and refined grain size. Nd–Nb co-doping enhanced the ferroelectric, dielectric, and magnetic properties. The primary leakage mechanism of the co-doped sample was based on combined SCLC and ohmic conduction. The $Bi_{0.9}Nd_{0.1}Fe_{0.95}Nb_{0.05}O_3$ ceramic sample exhibited optimal values of P_r and moderate M_r at ~3.12 μC cm^{-2} and 0.12 emu g^{-1}, respectively. The improvement in ferromagnetic and ferroelectric properties of the Nd–Nb co-doped BF samples was attributed to the crystal structure, decreased leakage current, and the presence of a disrupted spiral spin configuration. These findings may offer fresh insights for advancing the multiferroic properties of $BiFeO_3$ ceramics through compositional optimization and grain size manipulation, potentially paving the way for significant practical applications.

Author Contributions: Methodology, H.Y.; investigation, Z.Z. and M.W.; data curation, X.W., Y.C., and H.M.; writing—original draft, T.W. and P.R.; writing—review and editing, T.W.; supervision, S.S. and Q.M.; project administration, Q.M. All authors have read and agreed to the published version of the manuscript.

Funding: This work was supported by the Guangdong Basic and Applied Basic Research Foundation (grant No. 2022A1515140002 and 2023A1515140074), the College Students' Innovative Entrepreneurial Training Plan Program (S202310577064), the Project of General Colleges and Universities in Guangdong Province (grant No. 2021ZDJS079), the Program for Innovative Research Team of Guangdong Province & Huizhou University (IRTHZU), the Shenzhen Science and Technology Program (No.

JCYJ20220531094612029), and the National Natural Science Foundation of China (grant No. 62204162). The authors would like to express their gratitude to MJEditor (www.mjeditor.com, accessed on 30 December 2024) for the expert linguistic services.

Data Availability Statement: Data is contained within the article.

Conflicts of Interest: The authors declare no conflicts of interest.

References

1. Hu, Z.; Stenning, G.B.G.; Zhang, H.; Shi, Y.; Koval, V.; Hu, W.; Zhou, Z.; Jia, C.; Abrahams, I.; Yan, H. Non-volatile voltage-controlled magnetization in single-phase multiferroic ceramics at room temperature. *J. Mater.* **2025**, *11*, 100857. [CrossRef]
2. Sun, Y.; Liu, J.; Sun, T.; Yu, Z.; Zheng, Z.; Ge, M.; Bai, L. Structural evolution and enhanced multiferroicity in $BiFeO_3$-based ceramics via rare earth element co-substitution. *J. Eur. Ceram. Soc.* **2025**, *45*, 117021. [CrossRef]
3. Billah, A.; Matsuno, Y.; Anju, A.N.; Koike, K.; Kubota, S.; Hirose, F.; Ahmmad, B. Unusual behavior of magnetic coercive fields with temperature and applied field in La-doped $BiFeO_3$ ceramics. *ACS Appl. Electron. Mater.* **2023**, *5*, 4261–4267. [CrossRef]
4. Liu, L.; Zhu, X.L.; Chen, X.M. Improved ferroelectricity and magnetoelectric coupling effect of multielements co-substituted $BiFeO_3$ ceramics. *J. Am. Ceram. Soc.* **2023**, *107*, 933–944. [CrossRef]
5. Selbach, S.M.; Tybell, T.; Einarsrud, M.A.; Grande, T. The ferroic phase transitions of $BiFeO_3$. *Adv. Mater.* **2008**, *20*, 3692–3696. [CrossRef]
6. Catalan, G.; Scott, J.F. Physics and applications of bismuth ferrite. *Adv. Mater.* **2009**, *21*, 2463–2485. [CrossRef]
7. Gao, X.; Zhang, L.; Wang, C.; Shen, Q. Enhanced magnetic and ferroelectric properties of textured multiferroic $Bi_5Ti_3FeO_{15}$-$CoFe_2O_4$ ceramics prepared by plasma activated sintering. *Ceram. Int.* **2017**, *43*, 6607–6611. [CrossRef]
8. Fiebig, M. Revival of the magnetoelectric effect. *J. Phys. D Appl. Phys.* **2005**, *38*, R123–R152. [CrossRef]
9. Teague, J.R.; Gerson, R.; James, W.J. Dielectric hysteresis in single crystal $BiFeO_3$. *Solid. State Commun.* **1970**, *8*, 1073–1074. [CrossRef]
10. Shi, X.X.; Liu, X.Q.; Chen, X.M. Readdressing of magnetoelectric effect in bulk $BiFeO_3$. *Adv. Funct. Mater.* **2017**, *27*, 1604037. [CrossRef]
11. Boukhari, M.; Abdelkafi, Z.; Abdelmoula, N.; Khemakhem, H.; Randrianantoandro, N. Enhanced dielectric and optical properties in Zn^{2+} and Zr^{4+} co-doping BiFeO3 ceramic. *J. Mater. Sci. Mater. Electron.* **2023**, *34*, 1218. [CrossRef]
12. Lakshmi, S.D.; Banu, I.B.S. Multiferroism and magnetoelectric coupling in single-phase Yb and X (X = Nb, Mn, Mo) co-doped BiFeO3 ceramics. *J. Sol-Gel Sci. Technol.* **2018**, *89*, 713–721. [CrossRef]
13. Abe, K.; Sakai, N.; Takahashi, J.; Itoh, H.; Adachi, N.; Ota, T. Leakage current properties of cation-substituted $BiFeO_3$ ceramics. *Jpn. J. Appl. Phys.* **2010**, *49*, 09MB01. [CrossRef]
14. Yamamoto, H.; Kihara, T.; Oka, K.; Tokunaga, M.; Mibu, K.; Azuma, M. Spin structure change in co-substituted $BiFeO_3$. *J. Phys. Soc. Jpn.* **2016**, *85*, 064704. [CrossRef]
15. Quan, Z.; Liu, W.; Hu, H.; Xu, S.; Sebo, B.; Fang, G.; Li, M.; Zhao, X. Microstructure, electrical and magnetic properties of Ce-doped $BiFeO_3$ thin films. *J. Appl. Phys.* **2008**, *104*, 084106. [CrossRef]
16. Chen, J.; Xu, B.; Liu, X.Q.; Gao, T.T.; Bellaiche, L.; Chen, X.M. Symmetry modulation and enhanced multiferroic characteristics in $Bi_{1-x}Nd_xFeO_3$ ceramics. *Adv. Funct. Mater.* **2019**, *29*, 1806399. [CrossRef]
17. Chen, J.; Liu, L.; Zhu, X.L.; Gareeva, Z.; Zvezdin, A.; Chen, X.M. The involvement of Pna21 phase in the multiferroic characteristics of La/Lu co-substituted $BiFeO_3$ ceramics. *Appl. Phys. Lett.* **2021**, *119*, 112901. [CrossRef]
18. Chen, J.; Su, N.; Zhu, X.L.; Liu, X.Q.; Zhang, J.-X.; Chen, X.M. Electric-field-controlled magnetism due to field-induced transition of Pna21/R3c in $Bi_{1-x}Gd_xFeO_3$ ceramics. *J. Mater.* **2021**, *7*, 967–975. [CrossRef]
19. Liu, L.; Zhu, X.L.; Gareeva, Z.; Zvezdin, A.; Eiras, J.A.; Chen, X.M. Symmetry evolution and modulation of multiferroic characteristics in Bi1-xLaxFeO_3 ceramics. *Appl. Phys. Lett.* **2022**, *120*, 13904. [CrossRef]
20. Liu, L.; Chen, J.; Zhu, X.L.; Chen, X.M. Structure evidence of Pna21 phase and field-induced transition of Pna21/R3c in $Bi_{1-x}Sm_xFe_{0.99}Ti_{0.01}O_3$ ceramics. *Appl. Phys. Lett.* **2021**, *118*, 142904. [CrossRef]
21. Petkov, V.; Zafar, A.; Kenesei, P.; Shastri, S. Chemical compression and ferroic orders in La substituted $BiFeO_3$. *Phys. Rev. Mater.* **2023**, *7*, 054404. [CrossRef]
22. Ke, H.; Zhang, L.; Zhang, H.; Li, F.; Luo, H.; Cao, L.; Wang, W.; Wang, F.; Jia, D.; Zhou, Y. Electric/magnetic behaviors of Nd/Ti co-doped $BiFeO_3$ ceramics with morphotropic phase boundary. *Scr. Mater.* **2019**, *164*, 6–11. [CrossRef]
23. Gumiel, C.; Jardiel, T.; Bernardo, M.S.; Villanueva, P.G.; Urdiroz, U.; Cebollada, F.; Aragó, C.; Caballero, A.C.; Peiteado, M. Combination of structural and microstructural effects in the multiferroic response of Nd and Ti co-doped $BiFeO_3$ bulk ceramics. *Ceram. Int.* **2019**, *45*, 5276–5283. [CrossRef]
24. Wang, T.; Song, S.H.; Ma, Q.; Tan, M.L.; Chen, J.J. Highly improved multiferroic properties of Sm and Nb co-doped $BiFeO_3$ ceramics prepared by spark plasma sintering combined with sol-gel powders. *J. Alloys Compd.* **2019**, *795*, 60–68. [CrossRef]

25. Gao, X.; Li, Y.; Chen, J.; Yuan, C.; Zeng, M.; Zhang, A.; Gao, X.; Lu, X.; Li, Q.; Liu, J.-M. High energy storage performances of $Bi_{1-x}Sm_xFe_{0.95}Sc_{0.05}O_3$ lead-free ceramics synthesized by rapid hot press sintering. *J. Eur. Ceram. Soc.* **2019**, *39*, 2331–2338. [CrossRef]
26. Lin, F.; Yu, Q.; Deng, L.; Zhang, Z.; He, X.; Liu, A.; Shi, W. Effect of La/Cr codoping on structural transformation, leakage, dielectric and magnetic properties of $BiFeO_3$ ceramics. *J. Mater. Sci.* **2017**, *52*, 7118–7129. [CrossRef]
27. Wang, C.A.; Pang, H.Z.; Zhang, A.H.; Lu, X.B.; Gao, X.S.; Zeng, M.; Liu, J.M. Enhanced ferroelectric polarization and magnetization in $BiFe_{1-x}Sc_xO_3$ ceramics. *Mater. Res. Bull.* **2015**, *70*, 595–599. [CrossRef]
28. Perez-Maqueda, L.A.; Gil-Gonzalez, E.; Wassel, M.A.; Jha, S.K.; Perejon, A.; Charalambous, H.; Okasinski, J.; Sanchez-Jimenez, P.E.; Tsakalakos, T. Insight into the $BiFeO_3$ flash sintering process by in-situ energy dispersive X-ray diffraction (ED-XRD). *Ceram. Int.* **2019**, *45*, 2828–2834. [CrossRef]
29. Bernardo, M.S.; Jardiel, T.; Caballero, A.C.; Bram, M.; Gonzalez-Julian, J.; Peiteado, M. Electric current activated sintering (ECAS) of undoped and titanium-doped $BiFeO_3$ bulk ceramics with homogeneous microstructure. *J. Eur. Ceram. Soc.* **2019**, *39*, 2042–2049. [CrossRef]
30. Cai, W.; Gao, R.; Fu, C.; Yao, L.; Chen, G.; Deng, X.; Wang, Z.; Cao, X.; Wang, F. Microstructure, enhanced electric and magnetic properties of $Bi_{0.9}La_{0.1}FeO_3$ ceramics prepared by microwave sintering. *J. Alloys Compd.* **2019**, *774*, 61–68. [CrossRef]
31. Yu, C.; Viola, G.; Zhang, D.; Zhou, K.; Koval, V.; Mahajan, A.; Wilson, R.M.; Tarakina, N.V.; Abrahams, I.; Yan, H. Phase evolution and electrical behaviour of samarium-substituted bismuth ferrite ceramics. *J. Eur. Ceram. Soc.* **2018**, *38*, 1374–1380. [CrossRef]
32. Koval, V.; Skorvanek, I.; Durisin, J.; Viola, G.; Kovalcikova, A.; Svec, P.; Saksl, K.; Yan, H. Terbium-induced phase transitions and weak ferromagnetism in multiferroic bismuth ferrite ceramics. *J. Mater. Chem. C* **2017**, *5*, 2669–2685. [CrossRef]
33. Wang, T.; Song, S.H.; Wang, M.; Li, J.Q.; Ravi, M. Effect of annealing atmosphere on the structural and electrical properties of $BiFeO_3$ multiferroic ceramics prepared by sol–gel and spark plasma sintering techniques. *Ceram. Int.* **2016**, *42*, 7328–7335. [CrossRef]
34. Wang, T.; Xu, T.; Gao, S.; Song, S.H. Effect of Nd and Nb co-doping on the structural, magnetic and optical properties of multiferroic $BiFeO_3$ nanoparticles prepared by sol-gel method. *Ceram. Int.* **2017**, *43*, 4489–4495. [CrossRef]
35. Wang, T.; Wang, X.L.; Song, S.H.; Ma, Q. Effect of rare-earth Nd/Sm doping on the structural and multiferroic properties of $BiFeO_3$ ceramics prepared by spark plasma sintering. *Ceram. Int.* **2020**, *46*, 15228–15235. [CrossRef]
36. Zhang, S.X.; Wang, L.; Chen, Y.; Wang, D.L.; Yao, Y.B.; Ma, Y.W. Observation of room temperature saturated ferroelectric polarization in Dy substituted $BiFeO_3$ ceramics. *J. Appl. Phys.* **2012**, *111*, 074105. [CrossRef]
37. Khomchenko, V.A.; Paixao, J.A.; Shvartsman, V.V.; Borisov, P.; Kleemann, W.; Karpinsky, D.V.; Kholkin, A.L. Effect of Sm substitution on ferroelectric and magnetic properties of $BiFeO_3$. *Scr. Mater.* **2010**, *62*, 238–241. [CrossRef]
38. Islam, M.R.; Islam, M.S.; Zubair, M.A.; Usama, H.M.; Azam, M.S.; Sharif, A. Evidence of superparamagnetism and improved electrical properties in Ba and Ta co-doped $BiFeO_3$ ceramics. *J. Alloys Compd.* **2018**, *735*, 2584–2596. [CrossRef]
39. Durga Rao, T.; Raja Kandula, K.; Kumar, A.; Asthana, S. Improved magnetization and reduced leakage current in Sm and Sc co-substituted $BiFeO_3$. *J. Appl. Phys.* **2018**, *123*, 244104. [CrossRef]
40. Habib, M.; Tang, L.; Xue, G.; Rahman, A.; Kim, M.-H.; Lee, S.; Zhou, X.; Zhang, Y.; Zhang, D. Design and development of a new lead-free $BiFeO_3$-$BaTiO_3$ quenched ceramics for high piezoelectric strain performance. *Chem. Eng. J.* **2023**, *473*, 145387. [CrossRef]
41. Park, T.-J., Papaefthymiou, G.C.; Viescas, A.J.; Moodenbaugh, A.R.; Wong, S. Size-dependent magnetic properties of single-crystalline multiferroic $BiFeO_3$ nanoparticles. *Nano Lett.* **2007**, *7*, 766–772. [CrossRef] [PubMed]
42. Wu, X.; Luo, H.; Guo, H.; Liu, J.; Bai, Y.; Zhao, S. Tuning of multiferroic traits in $BiFeO_3$ ceramics by electronic structure. *Ceram. Int.* **2024**, *50*, 18853–18867. [CrossRef]
43. Wang, B.; Liu, W.; Zhao, T.L.; Peng, W.; Ci, P.H.; Dong, S.X. Promising lead-free $BiFeO_3$-$BaTiO_3$ ferroelectric ceramics: Optimization strategies and diverse device applications. *Progress Mater. Sci.* **2024**, *146*, 101333. [CrossRef]
44. Radojković, A.; Luković-Golić, D.; Orsini, N.J.; Nikolić, N.; Ćirković, J.; Lazarević, S.; Despotović, Ž. Evolution of ferroelectric and piezoelectric properties of $BiFeO_3$ ceramics doped with lanthanum and zirconium. *J. Alloys Compd.* **2024**, *1009*, 176901. [CrossRef]
45. Zhang, L.; Ke, H.; Zhang, H.; Li, F.; Zhao, J.; Luo, H.; Cao, L.; Zeng, G.; Li, X.; Wang, W.; et al. Effects of morphotropic phase boundary on the electric behavior of Er/Ti co-doped $BiFeO_3$ ceramics. *Scr. Mater.* **2019**, *158*, 71–76. [CrossRef]
46. Zhang, W.; Zhu, X.; Wang, L.; Xu, X.; Yao, Q.; Mao, W.; Li, X. Study on the magnetic and ferroelectric properties of $Bi_{0.95}Dy_{0.05}Fe_{0.95}M_{0.05}O_3$ (M=Mn, Co) ceramics. *J. Supercond. Nov. Magn.* **2017**, *30*, 3001–3005. [CrossRef]
47. Hua, H.; Bao, G.; Li, C.; Zhu, Y.; Yang, J.; Li, X. Effect of Ho, Mn co-doping on the structural, optical and ferroelectric properties of $BiFeO_3$ nanoparticles. *J. Mater. Sci. Mater. Electron.* **2017**, *28*, 17283–17287. [CrossRef]
48. Chen, X.; Wang, J.; Yuan, G.; Wu, D.; Liu, J.; Yin, J.; Liu, Z. Structure, ferroelectric and piezoelectric properties of multiferroic $Bi_{0.875}Sm_{0.125}FeO_3$ ceramics. *J. Alloys Compd.* **2012**, *541*, 173–176. [CrossRef]
49. Jeong, J.; Le, M.D.; Bourges, P.; Petit, S.; Furukawa, S.; Kim, S.A.; Lee, S.; Cheong, S.W.; Park, J.G. Temperature-dependent interplay of Dzyaloshinskii-Moriya interaction and single-ion anisotropy in multiferroic $BiFeO_3$. *Phys. Rev. Lett.* **2014**, *113*, 107202. [CrossRef]

50. Yuan, G.L.; Or, S.W.; Liu, J.M.; Liu, Z.G. Structural transformation and ferroelectromagnetic behavior in single-phase $Bi_{1-x}Nd_xFeO_3$ multiferroic ceramics. *Appl. Phys. Lett.* **2006**, *89*, 052905. [CrossRef]
51. Karpinsky, D.V.; Silibin, M.V.; Latushka, S.I.; Zhaludkevich, D.V.; Sikolenko, V.V.; Svetogorov, R.; Sayyed, M.I.; Almousa, N.; Trukhanov, A.; Trukhanov, S.; et al. Temperature-driven transformation of the crystal and magnetic structures of $BiFe_{0.7}Mn_{0.3}O_3$ ceramics. *Nanomaterials* **2022**, *12*, 2813. [CrossRef] [PubMed]
52. Abdel Monem, A.M.; Abd-Elmohssen, N.; El-Bahnasawy, H.H.; Makram, N.; Sedeek, K. Comparative studies by X-ray diffraction, Raman, vibrating sample magnetometer and Mössbauer spectroscopy of pure, Sr doped and Sr, Co co-doped $BiFeO_3$ ceramic synthesized via tartaric acid-assisted technique. *Ceram. Int.* **2023**, *49*, 15213–15220. [CrossRef]
53. Yadav, M.; Agarwal, A.; Sanghi, S.; Kotnala, R.K.; Shah, J.; Bhasin, T.; Tuteja, M.; Singh, J. Crystal structure refinement, dielectric and magnetic properties of A-site and B-site co-substituted $Bi_{0.90}Nd_{0.10}Fe_{1-x}Ti_xO_3$ (x=0.00, 0.02, 0.05 & 0.07) ceramics. *J. Alloys Compd.* **2018**, *750*, 848–856.

nanomaterials

Article

Interface Optimization and Thermal Conductivity of Cu/Diamond Composites by Spark Plasma Sintering Process

Junfeng Zhao *, Hao Su, Kai Li, Haijuan Mei *, Junliang Zhang and Weiping Gong *

Guangdong Provincial Key Laboratory of Electronic Functional Materials and Devices, Huizhou University, Huizhou 516001, China
* Correspondence: zhaojunfeng@hzu.edu.cn (J.Z.); haijuanmei@hzu.edu.cn (H.M.); gwp@hzu.edu.cn (W.G.)

Abstract: Cu/Diamond (Cu/Dia) composites are regarded as next-generation thermal dissipation materials and hold tremendous potential for use in future high-power electronic devices. The interface structure between the Cu matrix and the diamond has a significant impact on the thermophysical properties of the composite materials. In this study, Cu/Dia composite materials were fabricated using the Spark Plasma Sintering (SPS) process. The results indicate that the agglomeration of diamond particles decreases with increasing particle size and that a uniform distribution is achieved at 200 μm. With an increase in the sintering temperature, the interface bonding is first optimized and then weakened, with the optimal sintering temperature being 900 °C. The addition of Cr to the Cu matrix leads to the formation of Cr_7C_3 after sintering, which enhances the relative density and bonding strength at the interface, transitioning it from a physical bond to a metallurgical bond. Optimizing the diamond particle size increased the thermal conductivity from 310 W/m K to 386 W/m K, while further optimizing the interface led to a significant increase to 516 W/m K, representing an overall improvement of approximately 66%.

Keywords: Cu/diamond; interface structure; thermal conductivity

1. Introduction

At present, with the rapid development of electronic devices towards miniaturization, high power, high integration, and efficient heat dissipation are essential to maintain high performance and prolong equipment life [1–3]. Every 10 °C increase in the operating temperature of an electronic device reduces its reliability by half [4]. This significantly increases the power density of electronic devices, leading to a rapid increase in heat dissipation per unit area. How to transfer the large amount of heat generated has become the key to the safe, stable, and efficient operation of electronic devices, which brings new challenges to the current electronic packaging materials and technologies [5,6]. Materials that excel in thermal conductivity are essential for enhancing the heat dissipation of electronic devices. Materials with high thermal conductivity are urgently needed [7,8]. To address these challenges, metal matrix composites that boast superior thermal conductivity and tunable thermal expansion coefficients are regarded as the most promising thermal management materials for electronic packaging applications [9–11]. The thermal conductivity of Cu is approximately 385–401 W/(m·K), but it has a high coefficient of thermal expansion; the thermal conductivity of diamond exceeds 2000 W/mK, but its high cost, difficult processing, and extreme brittleness limit its applications. Due to its extremely high thermal conductivity, diamond is often used as a reinforcing phase to make metal-based electronic packaging materials. The thermal expansion coefficients of Cu/Dia composites can be

 https://doi.org/10.3390/nano15010073

changed by adjusting the component ratio, with significantly improved thermal conductivity, typically ranging from 400 to 1500 W/m·K, sometimes even higher. However, the smooth and chemically inert surface of diamond, combined with its poor wettability with the Cu matrix, leads to poor interface bonding and reduced thermal performance [12,13]. Therefore, optimizing the interface structure between the diamond and Cu matrix is crucial for improving the thermal conductivity of the metal matrix composites.

For metal matrix composites, the interface region plays a decisive role in the overall performance of the material. In the process of preparing composite materials, if the interface structure between the matrix and the reinforcement phase is not well controlled, the excellent properties of the composites will be greatly reduced [14]. A large body of literature shows that if the interface quality of diamond and Cu is poor, the thermal conductivity of the Cu/Dia composites not only becomes lower than that of copper but is also much lower than that of diamond with the addition of diamond [15–17]. Due to the significance and complexity of the interface in Cu/Dia composite materials, the optimal design of the interface has become a research topic of great concern to scholars [18,19]. The alloying of Cu matrix and strong carbon compounds can significantly improve the infiltration and interface bonding between Cu and diamond [20–22]. The strong carbon compounds and diamond form an interface layer, which increases the diffusion and reaction with the diamond surface and improves the wettability of Cu and diamond [23]. This interface can also promote the coupling of phonons and diamond phonons in the Cu matrix, thereby improving the thermal conductivity of the composite. Weber [24] found that the alloying elements Cr and B in the matrix have an effect on the thermal conductivity and thermal expansion coefficient of the composite. The research indicated that as the content of alloying elements in the matrix increases, the bonding at the two-phase interface is significantly enhanced. However, upon further increasing the alloying elements, an optimal content was identified where the thermal conductivity and thermal expansion coefficient of the composite material reached their peak. Both Bai and Chu et al. found that the thermal conductivity of Cu/Dia composites first increased and then decreased with the increase in Zr and B contents, and the maximum thermal conductivity was obtained at 1037 °C and 1.2 wt.% Zr content [11,25]. Kang et al. coated diamond particles with a W layer and prepared Cu/Dia (W) composites by pressure infiltration method [26]; the results showed that the W coating played an effective sticking role and had good interface modification. In the development of Cu/Cr/diamond composites, the latest research focuses on enhancing the thermal conductivity of the materials through precise composition adjustment and innovative fabrication techniques [27,28]. Specifically, the use of Cr alloying technology to improve the performance of the copper matrix, combined with Cr-coated diamond particles, effectively optimizes the interface structure, thereby significantly enhancing the thermal conductivity [29,30]. Additionally, the introduction of chromium compound layers and the construction of a three-dimensional diamond network create an efficient pathway for phonon transport, further improving the efficiency of heat conduction. To meet the heat dissipation requirements of high-power electronic devices, research is ongoing to optimize the particle size and fabrication process of the composites and to further improve the comprehensive thermomechanical properties of the materials by enhancing the interface bonding strength. Therefore, to further investigate the influence of alloying elements on the interface characteristics of Cu/Dia composites, this study focused on analyzing the effects of different Cr contents and various diamond particle sizes on the thermal conductivity of the composites, seeking to achieve the optimal alloying element ratio for the best thermal conductivity.

In this work, the optimal method for fabricating Cu/Dia composites with high thermal conductivity was explored. The Cu/Dia composites were prepared using the SPS technique with the optimized process, and the influence of Cr content on the interfacial products and thermal properties of the Cu/Dia composites was investigated. This study will establish a theoretical framework for enhancing the thermal conductivity of Cr-modified Cu/Diamond composites through the effective design and manipulation of the interface between Cu and diamond.

2. Experimental

2.1. Raw Materials

Commercial purity Cu powder (99.99% purity) was used as a matrix material, and the Cu powder utilized was produced by Shanghai Macklin Biochemical Co., Ltd. with the product code C805729. The diamond crystal grains were purchased from Huanghe Whirl wind company. The morphology of the raw materials is shown in Figure 1.

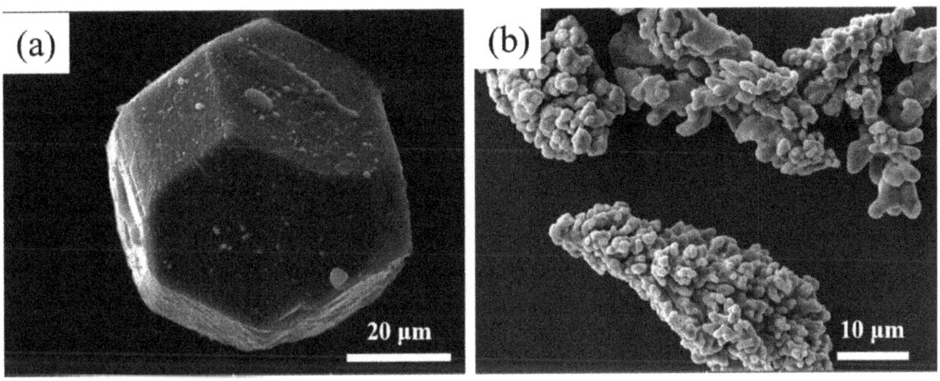

Figure 1. SEM micrographs of original diamond (**a**) and Cu powder (**b**).

2.2. Sample Preparation

The diamond particles, Cr powder, and Cu powder were weighed according to the set mass ratio and placed into a ball milling jar, followed by adding an appropriate amount of ball milling beads. The ball milling beads and the weighed powder were added to the jar in a ball-to-powder ratio of 10:1. To prevent the oxidation of Cu powder during the ball milling process, a vacuum pump was used to evacuate the air. The vacuum ball milling jar was then placed into a planetary ball mill and milled for 5 h. The grinding parameters were set as follows: rotation speed of 200 rpm, 15 min for forward rotation, and 15 min for reverse rotation with a 4 min interval. After ball milling was completed, the beads and powder were separated using a sieve. The obtained Cu/diamond powder was placed into a graphite mold, and the powder was then pressed into discs with a diameter of 13 mm at a pressure of 50 MPa and sintered under vacuum at the set temperature to form the shape. The sample preparation was carried out using a Spark Plasma Sintering (SPS) machine model SPS-12000-10AT (manufactured by Shenzhen Xingyuan Discharge Plasma Technology Co., Ltd., Shenzhen, China). During the sintering process, the sample was first heated to 500 °C at a rate of 50 °C/min; then, pressure was applied while continuing to heat to the set sintering temperature. After reaching the set temperature, it was maintained for 20 min to ensure full densification. After sintering was completed, the power was turned off and the sample was allowed to cool in the furnace. The pressure was slowly released when the temperature dropped to 200 °C. The diamond volume fraction amounted to approximately 40% of all samples. The Cu/Dia composites with this composition can effectively enhance thermal conductivity while avoiding excessive reduction in the material's strength and toughness.

2.3. Characterization

The composites obtained were of a cylinder geometry with diameter of 13 mm and thickness of 5 mm. The surface morphology and distribution of elements for the Cu/Dia samples were studied by scanning electron microscope (SEM, Tescan VEGA3 XMU, Brno, Czech Republic). The composition was confirmed via energy dispersive spectroscopy (EDS) with a 30 kV accelerating voltage. The thermal conductivity (λ) of the room temperature-sintered composites was determined using the thermal diffusivity (α), density (ρ), and specific heat capacity (C_p) values based on the following formula: $\lambda = \alpha \times C_p \times \rho$. The thermal diffusivities of Cu/Dia composites were measured by thermal conductivity tester (LFA467HyperFlash, Netzsch, Selb, Germany). The different phases within the composite material's structure were identified by X-ray diffraction (XRD, Rigaku miniflex600, Tokyo, Japan) using Cu Kα radiation. The Archimedes principle method was used to determine the density of the Cu/Dia composite materials with different compositions.

3. Results and Discussion

3.1. Effect of Particle Size on Agglomeration

Figure 2 illustrates the microstructure of Cu/Dia composites with varying particle sizes at 900 °C. It is observed that when the diamond particle size is small, a pronounced aggregation phenomenon occurs among the particles, as indicated by the red circle in the figures a and b. As the size of the diamond particles increases, the degree of aggregation gradually diminishes. As depicted in Figure 2, smaller particle sizes are more prone to aggregation during the fabrication process. This phenomenon can be attributed to the fact that diamond particles with smaller dimensions possess a larger specific surface area and higher surface energy. The higher surface energy indicates stronger interparticle attractions, leading the particles to cluster together in an effort to reduce the overall system energy. Consequently, the formation of small pores within the aggregates and larger pores between them collectively increases the material's porosity and results in a decrease in density. As the diamond particle size increases, the agglomeration gradually decreases, and when the particle size reaches 200 μm, the diamond particles are uniformly distributed within the Cu matrix.

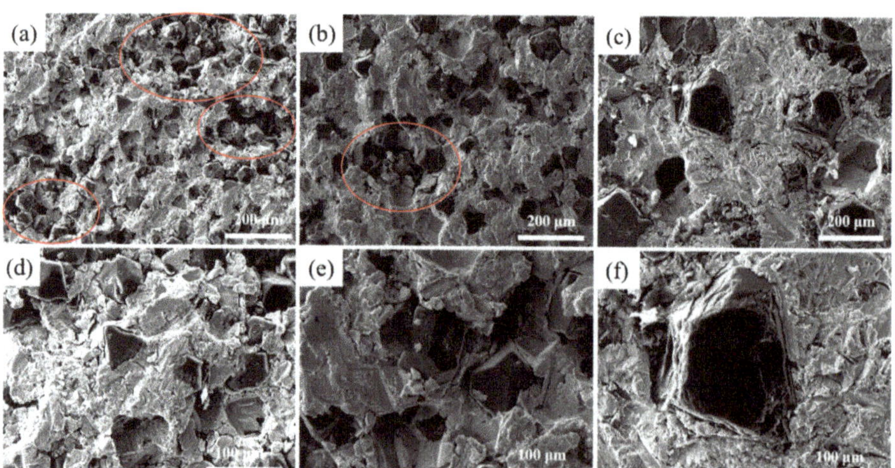

Figure 2. Microstructure of diamond/Al composites with different particle sizes. (**a**,**d**): 40 μm; (**b**,**e**): 80 μm; (**c**,**f**): 200 μm.

Figure 3 shows the effect of diamond particle sizes on the density and thermal properties of the composite materials at 900 °C. As the diamond particle size increases, both the relative density and actual density exhibit an upward trend. Particularly, when the diamond particle size increases from 40 μm to 200 μm, the relative density significantly increases from 92.8% to 95.01%. This indicates that larger diamond particles contribute to improving the overall density of the composite materials. With the increase in diamond particle size, the densification of the diamond/Cu composite material gradually improves. This might be because larger diamond particles are less prone to agglomeration in the composite materials, thereby reducing porosity and increasing the overall density of the materials. Agglomerated particles cannot be distributed uniformly in the matrix material, leading to non-uniform particle distribution throughout the composite materials and resulting in the formation of voids around the agglomerated areas. The agglomerated diamond particles hinder the flow of the matrix material, making it difficult for the matrix to fill the gaps between particles, thus increasing the porosity. The reduced interface bonding area and weakened bonding strength between the agglomerated particles and the matrix material also affect the overall densification of the composites. Additionally, within a given volume, as the particle size decreases, the number of interfaces increases, which in turn leads to a higher number of interfacial voids, resulting in a lower overall density.

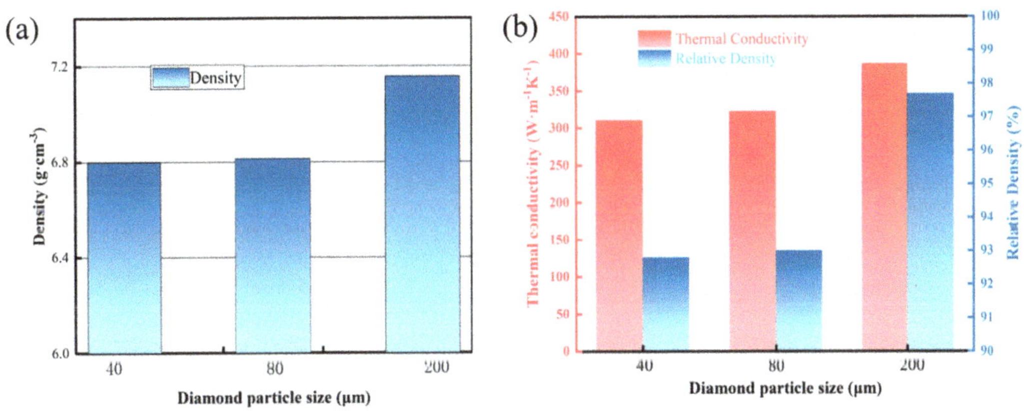

Figure 3. Density and thermal properties of composites with various particle sizes. (**a**): Relative density; (**b**): Thermal conductivity and actual density.

Figure 3b presents the thermal conductivity and actual density of composites with different particle sizes. The results indicate that the thermal conductivity of the composites gradually increases with the enlargement of the diamond particle diameter. When the diamond particle diameter reaches 200 μm, the thermal conductivity reaches its highest value. The uniform distribution of particles plays a positive role in enhancing the thermal conductivity of the composites. In Cu/Dia composites, the diamond particles act as excellent thermal conductors and are uniformly distributed, which helps to construct effective thermal conduction paths and thus improves the overall thermal conductivity. Conversely, particle agglomeration or non-uniform distribution can easily form defects within the material, such as voids and cracks, which may become areas of stress concentration and consequently weaken the material's performance. Uniformly distributed particles help to reduce these internal defects and optimize interface characteristics. The improvement in relative density can enhance the efficiency of heat transfer between particles, thereby enhancing the overall performance of the composites.

The standard theoretical models employed for predicting thermal conductivity are the Maxwell–Eucken model and the Discrete Element Method (DEM) model [31,32]. The prediction formula for the Maxwell–Eucken model is presented in Equation (1), while the DEM model's prediction formula is detailed in Equation (2) [31]:

$$\lambda = \frac{\lambda_{Cu}[2\lambda_{Cu} + \lambda_{diamond} + 2(\lambda_{diamod} - \lambda_{Cu})V_{diamond}]}{2\lambda_{Cu} + \lambda_{diamond} - (\lambda_{diamond} - \lambda_{Cu})V_{diamond}} \quad (1)$$

$$\sqrt[3]{\frac{\lambda}{\lambda_{Cu}}}(1 - V_{diamond}) = \frac{\frac{\lambda_{diamond}}{\lambda_{Cu}} - \frac{\lambda}{\lambda_{Cu}}}{\frac{\lambda_{diamond}}{\lambda_{Cu}} - 1} \quad (2)$$

In these equations, λ represents the thermal conductivity of the composite, λ_{Cu} denotes the thermal conductivity of the Cu matrix, $\lambda_{diamond}$ refers to the theoretical thermal conductivity of diamond, and $V_{diamond}$ is the volume fraction of diamond within the composite.

The thermal conductivity of the composites is influenced by the particle size of diamond, as it impacts the interface thermal resistance [33,34]. The modified effective thermal conductivity, $\lambda' d_{iamond}$, is utilized in place of the original $\lambda_{diamond}$.

$$\lambda\prime = \frac{\lambda_{diamond}}{1 + \frac{\lambda_{diamond}}{rh_c}} \quad (3)$$

In the above equation, r denotes the size of the diamond particle, and h_c represents the thermal conductivity at the interface between the Cu matrix and the diamond particles. As the diamond particle size decreases, the interface area increases, which diminishes phonon scattering and enhances phonon-mediated heat transfer. Consequently, when the volume fraction of diamond remains constant, a reduction in particle size leads to a decrease in the thermal conductivity. Conversely, as the size of the diamond particles increases and the interfacial gap diminishes, the thermal conductivity of the composite progressively improves.

3.2. Effect of Different Sintering Temperatures on the Interface

Figure 4 shows the effect of different sintering temperatures on the interface structure of Cu/Dia composites. As the temperature increases, the interface bonding condition improves under a certain pressure. However, when the temperature reaches 1000 °C, noticeable gaps appear at the interface. At an appropriate sintering temperature, Cu and diamond can form good chemical bonds, thereby enhancing the interface bonding strength. If the sintering temperature is too low, the interface reaction will be insufficient, leading to weaker bonding forces. Conversely, if the sintering temperature is too high, due to the difference in thermal expansion coefficients between Cu and diamond, a significant expansion discrepancy will occur at high temperatures, resulting in the formation of tiny gaps at the interface upon cooling. This reduces the wettability of Cu on diamond and subsequently affects the interface bonding effectiveness.

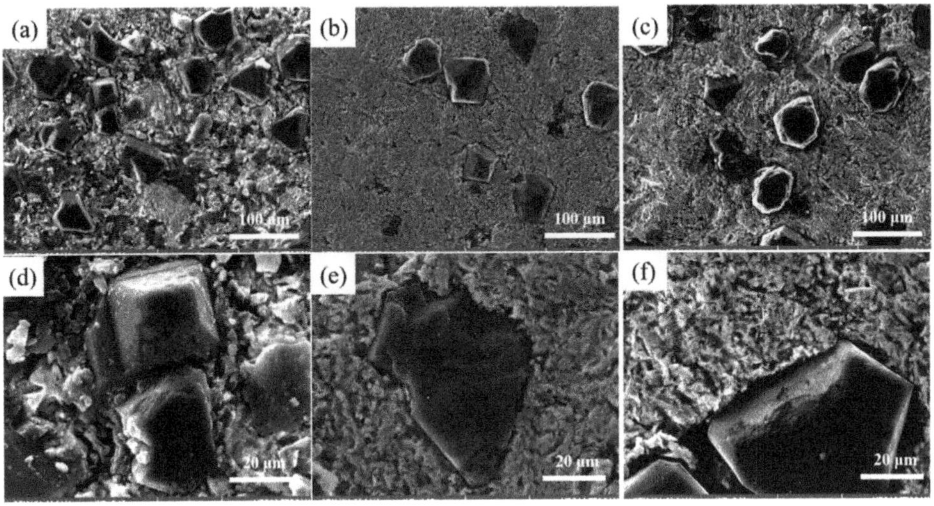

Figure 4. The impact of different sintering temperatures on the interface structure. (**a**,**d**) 800; (**b**,**e**) 900; (**c**,**f**) 1000 °C.

3.3. The Impact of Different Cr Contents on the Interface

Figure 5 illustrates the effect of different Cr contents on the interface structure of Cu/Dia composites at 900 °C. Figure 6 provides the XRD analysis of composites with various Cr contents. These composites consist of three distinct regions: the diamond reinforcement phase, the interface layer, and the Cu matrix, with the diamond reinforcement phase connected to the Cu matrix through the interface layer. At the interface without Cr, the bonding force between the diamond particles and pure Cu is weak, as shown in Figure 5a,d. It is clearly evident from the figures that despite the application of a certain pressure during manufacturing, there is a noticeable gap between the two phases due to the poor wettability between diamond and Cu, resulting in a lower thermal conductivity of the Cu/Dia composites [35]. Figure 5b,e shows the image of a Cu/Dia composite containing 1 wt% Cr; the bonding between diamond and Cu is significantly stronger than that in the previous interface. With the increase in Cr content in the Cu alloy, a transition from weak to strong interface bonding can be observed in the Cu/Dia composites. A small amount of Cr in the matrix is difficult to form a uniform interface due to insufficient chemical reaction. When the Cr content is increased to a sufficient value (3%), it can be confirmed that a strong interface bond is formed between the diamond particles and the Cu matrix, as shown in Figure 5c,f.

From Figure 6, the XRD of composites with different Cr contents reveals that the interface reaction product is Cr_7C_3. At the interface, the Cr element has undergone a chemical reaction with the diamond, leading to an improvement of the interface. It can be observed that as the Cr content increases from 0% to 3%, the constituent phases at the interface become tightly bonded, indicating that the interfacial carbide layer can act as a "binder" to firmly adhere the diamond filler to the Cu matrix. The Cr element reacts chemically with the diamond to form a layer of chromium carbide, which serves as a transition layer, enhancing the chemical bonding force between Cu and diamond. Extensive literature research indicates that Cu and diamond have poor wettability [22,36]; the addition of Cr promotes the diffusion of Cu atoms on the diamond surface, and this diffusion of Cu atoms on the diamond surface aids in the formation of a stronger interfacial bond.

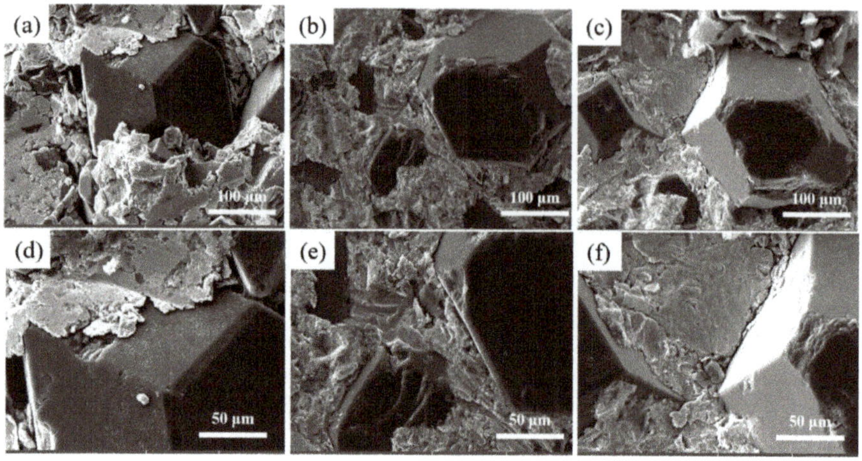

Figure 5. The influence of different Cr contents on the interface. (**a**,**d**) 0; (**b**,**e**) 1%; (**c**,**f**) 3%.

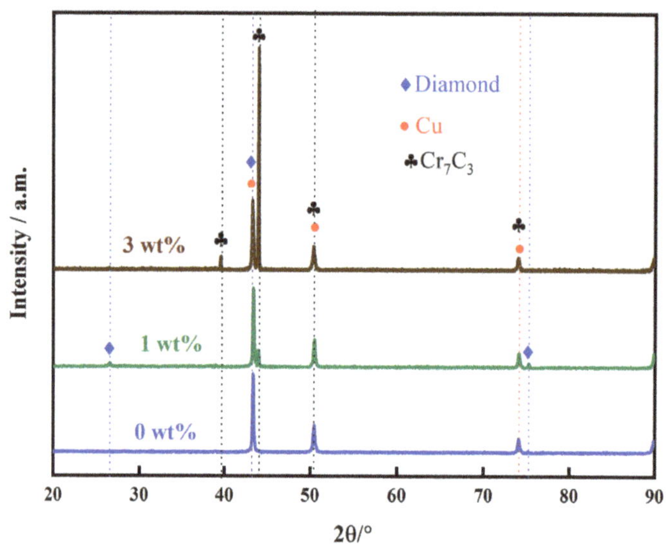

Figure 6. XRD of composite materials with different Cr contents.

Figure 7 presents the density and thermal properties of composites with varying Cr contents. When Cu and diamond are in direct contact, the presence of numerous voids at the interface results in a thermal conductivity of only 393 W/m K. As the interface layer is optimized, both the density and compactness of the composites increase, leading to a rising trend in the thermal conductivity coefficient. When the Cr content reaches 3%, the thermal conductivity attains its maximum value of 516 W/m K. The relationship between thermal conductivity and relative density indicates that as the interfacial voids decrease, both the thermal conductivity and relative density of the Cu/Dia composites gradually improve. This is because samples with lower relative density contain a higher volume of pores, and the thermal conductivity of gases is very poor, at just 0.0267 W/m K, which is negligible compared to the Cu matrix and diamond. Moreover, as the relative density decreases, the phonon scattering intensifies, reducing the mean free path of phonons and thus decreasing the thermal conductivity of the composites. This experimental result also fully demonstrates the significant role that relative density plays in the thermal conductivity of composite materials.

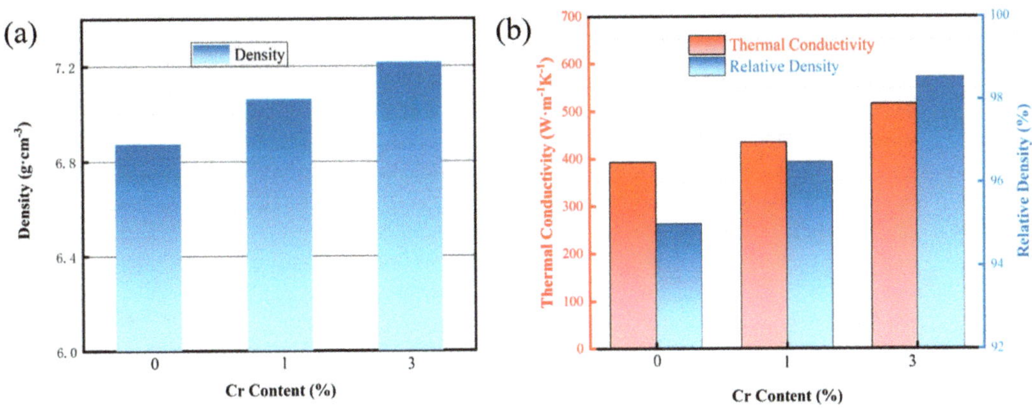

Figure 7. Density and thermal properties of composites with various Cr contents. (**a**) Density; (**b**) Thermal conductivity and Relative density.

When calculating the thermal conductivity of composite materials using a thermal conductivity prediction model, the interfacial thermal conductivity h_c between Cu and diamond is introduced, and its expression is as follows:

$$h_c = \frac{1}{4}\rho_{Cu}c_{Cu}C_{Cu}\eta_{Interface} \tag{4}$$

In the formula, ρ_{Cu} stands for the density of Cu, c_{Cu} is the specific heat capacity of copper, C_{Cu} indicates the phonon propagation speed in Cu, and $\eta_{Interface}$ represents the phonon heat dissipation coefficient at the interface of Cu and diamond. This coefficient can be calculated using the phonon mismatch model, as specified in Equation (5):

$$\eta_{Interface} = \frac{2Z_{Cu}Z_{diamond}}{(Z_{Cu}+Z_{diamond})^2}\frac{C_{DCu}}{C_{Ddiamond}} \tag{5}$$

Z_{Cu} and $Z_{diamond}$ are the respective phonon impedances of the Cu matrix and the diamond ($Z = \rho C_D$, C_D is the Debye sound velocity). $C_{Ddiamond}$ denotes the phonon velocity of diamond. The phonon velocities for both the Cu matrix and the diamond are determined using Equation (6):

$$C_D = \frac{1}{\sqrt{\frac{1}{2}\left(\frac{1}{C_l^2}+\frac{1}{C_t^2}\right)}} \tag{6}$$

C_l represents the phonon velocity traveling through the matrix in the longitudinal direction, while C_t is the phonon velocity propagating laterally within the matrix. The heat transfer in the Cu/Dia composites primarily relies on the phonons generated in the diamond. The interface and the pores are the key factors that influence thermal conduction in these composites [23]. The relationship between the thermal conductivity of Cu/Dia composite materials and the thickness of the interfacial layer can be explained by the following reasons: The absence or minimal addition of Cr content results in direct contact between the diamond and Cu, forming a discontinuous interface. Defects or voids at the interface increase the interfacial thermal resistance, significantly reducing the thermal conductivity of the Cu/Dia composites. The thermal conductivity improves with the optimization of the interface. Without the intervention of Cr alloy compounds at the interface, there are numerous voids present. As the Cr alloy compounds continuously distribute within the interfacial layer, the thermal conductivity of the Cu/Dia composites begins to show an upward trend. This indicates that interfacial modification of the composites through Cr alloying can introduce an interfacial layer between the Cu matrix and the diamond reinforcement phase, thereby improving the interfacial bonding and enhancing the thermal conductivity of the Cu/Dia composites. The role of the interfacial layer is mainly reflected in two aspects: first, it increases the wettability between the Cu matrix and the diamond, improving the interfacial quality of the composites and significantly reducing the interfacial thermal resistance; second, the acoustic properties of the Cr compounds are intermediate between those of Cu and diamond, which can increase the efficiency of phonon transmission, thus effectively enhancing the interfacial thermal conductivity. However, as the Cr content at the interface continues to increase, the thermal conductivity will exhibit a decreasing trend. On one hand, a certain amount of Cr_7C_3 is formed at the interface between the diamond and Cu, which improves the wettability and interfacial bonding strength between the matrix and the diamond, allowing the Cu/Dia composites to fully utilize the high thermal conductivity advantage of diamond. On the other hand, the thermal conductivity of Cr_7C_3 is lower than that of Cu and diamond, with only 19.1 W/m K. An excessive content of Cr_7C_3 will increase the interfacial thermal resistance, thereby reducing the thermal conductivity of the composites [37]. Therefore, strict control of interfacial products is necessary, as they are influenced by temperature and time. Future work will focus on optimizing the Cr diffusion layer thickness to boost the thermal conductivity of composites.

Compared with those reported in the literature for similar Cu/Dia composites, the results showed that after optimizing the interface structure, the thermal conductivity of our composites reached 516 W/mK, which is lower than some of the highest values reported (810 W/mK [38] and 696 W/mK [39]). This may be due to factors such as the lower diamond content, interface quality, particle size, and sintering temperature. Although our results are slightly below some of the highest values reported in the literature, the outcome remains competitive given the lower diamond content and reduced sintering temperature of our composites. Furthermore, our research, which focuses on enhancing thermal conductivity through interface optimization, offers valuable insights for the development of Cu/Dia composites with superior thermal performance.

4. Conclusions

Cu/Dia composites with Cr addition were fabricated using the SPS technology. The SPS process was refined, and the interfacial structure and thermal conductivity were assessed. Key findings are summarized below:

1. As the diamond particle size increases, the agglomeration of the composites gradually diminishes, and when the diamond particle size reaches 200 μm, the diamonds are uniformly distributed within the matrix. With the increase in sintering temperature, the interface bonding is optimized initially and then weakened, with the optimal sintering temperature being 900 °C.
2. The addition of Cr elements to the Cu matrix leads to the formation of the Cr_7C_3 phase after sintering, which increases the relative density of the composite material and enhances the bonding strength between the diamond and the matrix, transitioning the interface from a physical bond to a metallurgical bond.
3. With the optimization of diamond particle size, the thermal conductivity of the composites increased from 310 to 386 W/m K; after interface optimization, the thermal conductivity further rose to 516 W/m K, with an increase of approximately 66%.

Author Contributions: Formal analysis, J.Z. (Junliang Zhang); Investigation, H.S.; Resources, K.L.; Data curation, H.M.; Writing—original draft, J.Z. (Junfeng Zhao); Project administration, W.G. All authors have read and agreed to the published version of the manuscript.

Funding: This study was financially supported by the Featured Innovation Project of General Colleges and Universities in Guangdong Province (2022KTSCX137) and Guangdong Province Key Discipline Research Capacity Enhancement Project (2021ZDJS079).

Data Availability Statement: The original contributions presented in this study are included in the article. Further inquiries can be directed to the corresponding authors.

Conflicts of Interest: The authors declare no conflict of interest.

References

1. Yang, Z.; Albrow-Owen, T.; Cai, W.; Hasan, T. Miniaturization of optical spectrometers. *Science* **2021**, *371*, eabe0722. [CrossRef]
2. Zhao, J.; Lu, H.; Zhao, X.; Malyi, O.I.; Peng, J.; Lu, C.; Li, X.; Zhang, Y.; Zeng, Z.; Xing, G.; et al. Printable ink design towards customizable miniaturized energy storage devices. *ACS Mater. Lett.* **2020**, *2*, 1041–1056. [CrossRef]
3. Mathew, J.; Krishnan, S. A review on transient thermal management of electronic devices. *J. Electron. Packag.* **2022**, *144*, 010801. [CrossRef]
4. Bajenescu, T.I.; Bazu, M.I. *Reliability of Electronic Components: A Practical Guide to Electronic Systems Manufacturing*; Springer Science & Business Media: Berlin/Heidelberg, Germany, 2012.
5. Wang, Z.-G.; Jin, Y.-F.; Hong, R.; Du, J.; Dai, K.; Zhang, G.-Q.; Gao, J.; Xu, L.; Xu, J.-Z.; Li, Z.-M. Dual-functional thermal management materials for highly thermal conduction and effectively heat generation. *Compos. Part B Eng.* **2022**, *242*, 110084. [CrossRef]
6. Tang, D.-S.; Cao, B.-Y. Phonon thermal transport and its tunability in GaN for near-junction thermal management of electronics: A review. *Int. J. Heat Mass Transf.* **2023**, *200*, 123497. [CrossRef]
7. Moreno, G.; Narumanchi, S.; Feng, X.; Anschel, P.; Myers, S.; Keller, P. Electric-drive vehicle power electronics thermal management: Current status, challenges, and future directions. *J. Electron. Packag.* **2022**, *144*, 011004. [CrossRef]
8. Lv, Y.-G.; Wang, Y.-T.; Meng, T.; Wang, Q.-W.; Chu, W.-X. Review on thermal management technologies for electronics in spacecraft environment. *Energy Storage Sav.* **2024**, *3*, 153–189. [CrossRef]
9. Sharma, S.K.; Saxena, K.K.; Salem, K.H.; Mohammed, K.A.; Singh, R.; Prakash, C. Effects of various fabrication techniques on the mechanical characteristics of metal matrix composites: A review. *Adv. Mater. Process. Technol.* **2024**, *10*, 277–294. [CrossRef]
10. Maurya, P.; Kota, N.; Gibmeier, J.; Wanner, A.; Roy, S. Review on study of internal load transfer in metal matrix composites using diffraction techniques. *Mater. Sci. Eng. A* **2022**, *840*, 142973. [CrossRef]
11. Bai, H.; Ma, N.; Lang, J.; Zhu, C.; Ma, Y. Thermal conductivity of Cu/diamond composites prepared by a new pretreatment of diamond powder. *Compos. Part B Eng.* **2013**, *52*, 182–186. [CrossRef]
12. Dai, S.; Li, J.; Lu, N. Research progress of diamond/copper composites with high thermal conductivity. *Diam. Relat. Mater.* **2020**, *108*, 107993. [CrossRef]
13. Jia, S.; Yang, F. High thermal conductive copper/diamond composites: State of the art. *J. Mater. Sci.* **2021**, *56*, 2241–2274. [CrossRef]
14. Chen, L.; Huang, Z.; Kumar, S. Phonon transmission and thermal conductance across graphene/Cu interface. *Appl. Phys. Lett.* **2013**, *103*, 123110. [CrossRef]
15. Yoshida, K.; Morigami, H. Thermal properties of diamond/copper composite material. *Microelectron. Reliab.* **2004**, *44*, 303–308. [CrossRef]
16. Ekimov, E.; Suetin, N.; Popovich, A.; Ralchenko, V. Thermal conductivity of diamond composites sintered under high pressures. *Diam. Relat. Mater.* **2008**, *17*, 838–843. [CrossRef]
17. Bai, G.; Wang, L.; Zhang, Y.; Wang, X.; Wang, J.; Kim, M.J.; Zhang, H. Tailoring interface structure and enhancing thermal conductivity of Cu/diamond composites by alloying boron to the Cu matrix. *Mater. Charact.* **2019**, *152*, 265–275. [CrossRef]
18. Li, M.; Huang, J.; Fang, A.; Mansoor, B.; Pei, Z.; Ma, C. Binder jetting additive manufacturing of copper/diamond composites: An experimental study. *J. Manuf. Process.* **2021**, *70*, 205–213. [CrossRef]
19. Wu, K.; Zhang, L.; Li, F.; Sang, L.; Liao, M.; Tang, K.; Ye, J.; Gu, S. Enhancement of interfacial thermal conductance by introducing carbon vacancy at the Cu/diamond interface. *Carbon* **2024**, *223*, 119021. [CrossRef]
20. Zhang, Y.; Pershin, L.; Yang, Z.; Zhang, Y.; Hao, J.; Mostaghimi, J.; Zhang, H. Atmospheric plasma sprayed Cu coating on Cu-B/diamond composite for electronic packaging application. *Vacuum* **2024**, *228*, 113469. [CrossRef]
21. Li, H.; Li, K.; Fan, Y.; Liu, C.; Wang, C. Influence of brazing temperature on interfacial reaction layer characteristics of Cu-Sn-Ti/diamond composites. *Diam. Relat. Mater.* **2022**, *128*, 109276. [CrossRef]

22. Sang, J.; Yuan, Y.; Yang, W.; Zhu, J.; Fu, L.; Li, D.; Zhou, L. Exploring the underlying causes of optimizing thermal conductivity of copper/diamond composites by interface thickness. *J. Alloys Compd.* **2022**, *891*, 161777. [CrossRef]
23. Li, Q.; Liu, F.; Hu, S.; Song, H.; Yang, S.; Jiang, H.; Wang, T.; Koh, Y.K.; Zhao, C.; Kang, F.; et al. Inelastic phonon transport across atomically sharp metal/semiconductor interfaces. *Nat. Commun.* **2022**, *13*, 4901. [CrossRef]
24. Weber, L.; Tavangar, R. On the influence of active element content on the thermal conductivity and thermal expansion of Cu-X (X = Cr, B) diamond composites. *Scr. Mater.* **2007**, *57*, 988–991. [CrossRef]
25. Ke, C.; Jia, C.; Hong, G.; Li, W. On the thermal conductivity of Cu–Zr/diamond composites. *Mater. Des.* **2013**, *45*, 36–42.
26. Kang, Q.; He, X.; Ren, S.; Liu, T.; Liu, Q.; Wu, M.; Qu, X. Microstructure and thermal properties of copper–diamond composites with tungsten carbide coating on diamond particles. *Mater. Charact.* **2015**, *105*, 18–23. [CrossRef]
27. Chen, W.; Wang, F.; Fan, L.; Zheng, H.; Guo, X.; Zheng, P.; Zheng, L.; Zhang, Y. Double layer interfacial structure of Cr3C2–Cr7C3 in copper/diamond composites for thermal management applications. *Appl. Therm. Eng.* **2024**, *255*, 123958. [CrossRef]
28. Ukhina, A.V.; Dudina, D.V.; Esikov, M.A.; Samoshkin, D.A.; Stankus, S.V. The Influence of the Carbide-Forming Metallic Additives (W, Mo, Cr, Ti) on the Microstructure and Thermal Conductivity of Copper–Diamond Composites. *J. Compos. Sci.* **2023**, *7*, 219. [CrossRef]
29. Kang, Q.; He, X.; Ren, S.; Zhang, L.; Wu, M.; Guo, C.; Cui, W.; Qu, X. Preparation of copper–diamond composites with chromium carbide coatings on diamond particles for heat sink applications. *Appl. Therm. Eng.* **2013**, *60*, 423–429. [CrossRef]
30. Zhou, Y.; Hu, D.; Chen, M.; Wu, T.; Ouyang, J.; Xiong, D. An Investigation on the Spark Plasma Sintering Diffusion Bonding of Diamond/Cu Composites with a Cr Interlayer. *Materials* **2024**, *17*, 6026. [CrossRef]
31. Tong, X.C. *Advanced Materials for Thermal Management of Electronic Packaging*; Springer Science & Business Media: Berlin/Heidelberg, Germany, 2011; Volume 30.
32. Jiang, G.; Diao, L.; Kuang, K. *Advanced Thermal Management Materials*; Springer Science & Business Media: Berlin/Heidelberg, Germany, 2012.
33. Cho, H.J.; Yan, D.; Tam, J.; Erb, U. Effects of diamond particle size on the formation of copper matrix and the thermal transport properties in electrodeposited copper-diamond composite materials. *J. Alloys Compd.* **2019**, *791*, 1128–1137. [CrossRef]
34. Wu, X.; Luo, T. The importance of anharmonicity in thermal transport across solid-solid interfaces. *J. Appl. Phys.* **2014**, *115*, 014901. [CrossRef]
35. Huang, H.; Zhong, Y.; Cai, B.; Wang, J.; Liu, Z.; Peng, Q. Size-and temperature-dependent thermal transport across a Cu–diamond interface: Non-equilibrium molecular dynamics simulations. *Surf. Interfaces* **2023**, *37*, 102736. [CrossRef]
36. Liu, Z.; Zheng, S.; Lu, Z.; Pu, J.; Zhang, G. Adhesive transfer at copper/diamond interface and adhesion reduction mechanism with fluorine passivation: A first-principles study. *Carbon* **2018**, *127*, 548–556. [CrossRef]
37. Liu, X.; Sun, F.; Wang, L.; Wu, Z.; Wang, X.; Wang, J.; Kim, M.J.; Zhang, H. The role of Cr interlayer in determining interfacial thermal conductance between Cu and diamond. *Appl. Surf. Sci.* **2020**, *515*, 146046. [CrossRef]
38. Wang, L.; Li, J.; Che, Z.; Wang, X.; Zhang, H.; Wang, J.; Kim, M.J. Combining Cr pre-coating and Cr alloying to improve the thermal conductivity of diamond particles reinforced Cu matrix composites. *J. Alloys Compd.* **2018**, *749*, 1098–1105. [CrossRef]
39. Xie, Z.; Guo, H.; Zhang, X.; Huang, S.; Xie, H.; Mi, X. Tailoring the thermal and mechanical properties of diamond/Cu composites by interface regulation of Cr alloying. *Diam. Relat. Mater.* **2021**, *114*, 108309. [CrossRef]

nanomaterials

MDPI

Article

A Hierarchical Core-Shell Structure of NiO@Cu_2O-CF for Effective Non-Enzymatic Electrochemical Glucose Detection

Yueyun Huang, Jiahua You, Yingru Ding, Yun Xie, Ting Wang, Fanglong Zhu, Weiping Gong and Zhenting Zhao *

Guangdong Provincial Key Laboratory of Electronic Functional Materials and Devices, Huizhou University, Huizhou 516001, China; huangyy@hzu.edu.cn (Y.H.); 2207080116@stu.hzu.edu.cn (J.Y.); 2207080138@stu.hzu.edu.cn (Y.D.); xieyunxx@hzu.edu.cn (Y.X.); witton.wang@hzu.edu.cn (T.W.); flz_123@hzu.edu.cn (F.Z.); gwp@hzu.edu.cn (W.G.)

* Correspondence: zhzhting@hzu.edu.cn

Abstract: Non-enzymatic glucose detection is an effective strategy to control the blood glucose level of diabetic patients. A novel hierarchical core–shell structure of nickel hydroxide shell coated copper hydroxide core based on copper foam ($Ni(OH)_2$@$Cu(OH)_2$-CF) was fabricated and derived from NiO@Cu_2O-CF for glucose sensing. Cyclic voltammetry and amperometry experiments have demonstrated the efficient electrochemical catalysis of glucose under alkaline conditions. The measurement displays that the fabricated sensor exhibits a detection scale of 0.005–4.5 mM with a detection sensitivity of 4.67 μA/μM/cm^2. It has remarkable response/recovery times in respect of 750 μM glucose (1.0 s/3.5 s). Moreover, the NiO@Cu_2O-CF shows significant selectivity, reliable reproducibility and long-term stability for glucose determination, suggesting it is a suitable candidate for further applications.

Keywords: NiO-coated Cu_2O; non-enzymatic; electrochemical glucose detection

1. Introduction

Diabetes is a metabolic disease caused by the abnormal levels of glucose concentration in the human body. Nowadays, it is a serious chronic disease in the world. Diabetes is not only highly responsible for blindness, kidney failure, heart diseases, and strokes, but is also the leading cause of death [1–4]. "Silent killer" is another name for diabetes [5]. Diabetes was a direct cause of 1.5 million deaths in 2019 [6]. According to the World Health Organization (WHO), diabetes will become the seventh leading cause of death in the world by 2030 [3,7]. Considering the extremely large financial burden of diabetes and its serious complications, it is important to diagnosis diabetes early [6]. Monitoring glucose concentration is an effective technique in disease prevention and diagnosis.

Considerable efforts have been invested in the development of glucose determination, such as fluorescence [8–10], colorimetry [11,12], chemiluminescence [13,14], electrochemiluminescence [15] and the electrochemical technique [7,16–18]. Among these detecting methods, electrochemistry has been recognized as the promising approach for its inherent advantages such as convenience, rapid response, reliablity, outstanding sensitivity and the simplicity of operation. The electrochemical glucose sensors are generally divided into two major types, which are enzymatic and non-enzymatic methods [19]. Enzymatic glucose sensors, based on immobilizing glucolase on conductive substrates, possess high selectivity. However, the application of enzyme-based biosensors is limited by a lot of deficiencies, such as it being complicated to structure the electrodes, the low thermal and chemical stability of the enzyme, and it being expensive. Therefore, increasing studies focus on non-enzymatic glucose sensors [20–30]. Among a wide range of enzyme-free electrocatalytic materials, metal oxides, such as nickel oxide [31–33] and copper oxide [34–38], are suitable choices because they have the advantages of easy access and low cost.

To further enhance the electro-catalytic property of metal-oxide-based sensitive materials, preparing a three-dimensional multilayer structure is an effective approach [23,39,40]. In order to construct hierarchically sensitive materials, using copper foam (CF) as a substrate is an appropriate candidate. Furthermore, to increase the load of sensitive materials on a limited surface, designing a core-shell nano-structure is an excellent solution. Therefore, using nickel oxide and copper oxide to construct a core–shell structure on 3D copper foam and form a multicomponent hierarchical sensitive material could be a promising approach. It will have the advantages of a large specific surface area, efficient ion diffusion and electron transfer path.

This research drives the development of a flower spike-like structure composed of NiO and Cu_2O, formed by the air annealing of $Ni(OH)_2$ and $Cu(OH)_2$ based on copper foam, labeled as NiO@Cu_2O-CF. Within the newly created hierarchical architecture, Cu_2O nanowires spontaneously formed on the CF and were subsequently encapsulated by NiO nanosheets. In the hierarchical core–shell structure design, the prevention of nickel oxide accumulation ensures optimal contact with glucose, thereby enhancing sensor efficiency. The electrocatalytic property of the constructed sensor was investigated by cyclic voltammetry and amperometry tests. The as-prepared sensor possesses excellent performance in terms of high sensitivity, effective selectivity, a fast response–recovery characteristic and a large liner range for glucose sensing. Therefore, the NiO@Cu_2O-CF hierarchical structure is a promising material for sensitive glucose determination.

2. Materials and Methods

2.1. Materials and Reagents

Commercial Copper foams (CF) were purchased from Kunshan Luchuang ElectronicTechnology Co., Ltd. (Kunshan, China). All the reagents were obtained from Shanghai Aladdin Biochemical Technology Co., Ltd. (Shanghai, China) at analytical grade: ammonium persulfate ($(NH_4)_2S_2O_8$), sodium hydroxide (NaOH, 96%), concentrated hydrochloric acid (HCl), Nickel nitrate hexahydrate ($Ni(NO_3)_2 \cdot 6H_2O$, 97%), potassium chloride (KCl), sodium chloride (NaCl), fructose (Fru), sucrose (Suc), urea (UA, 99.0%), citric acid (CA) and sodium citrate (SSC, 98%).

2.2. Synthesis of NiO@Cu_2O-CF

The synthetic procedure for NiO@Cu_2O-CF is illustrated in Scheme 1 [25]. The illustration of the synthetic procedure and the photograph of the as-synthesized materials are shown in Scheme 1. Firstly, the copper foam substrate was soaked in 0.1 M HCl for 10 min, and then it was cleaned ultrasonically using DI water three times. Secondly, 16 mL 0.16 mol NaOH water solution and 44 mL 8 mmol $(NH_4)_2S_2O_8$ water solution were mixed and a colorless solution was obtained. Subsequently, CF was placed in the mixed solution for 15 min. Then, a obtained blue substrate was cleaned by absolute ethanol and DI water three times, which was marked as $Cu(OH)_2$-CF.

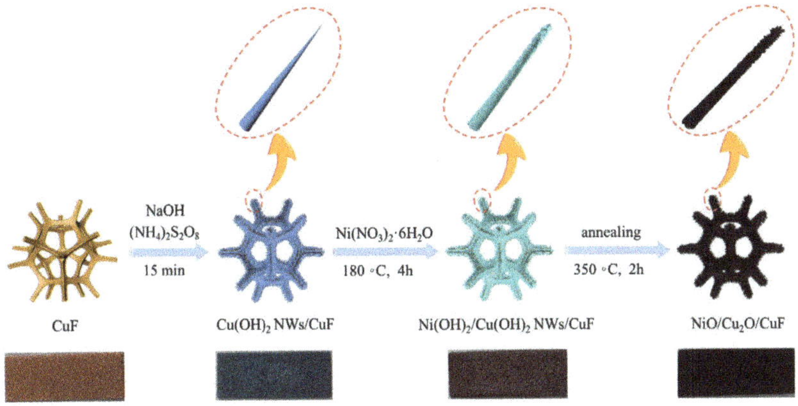

Scheme 1. The illustration of the synthetic procedure and the photographs of the as-synthesized materials.

The $NiO@Cu_2O$-CF electrode was derived by $Ni(OH)_2@Cu(OH)_2$-CF annealing in air. Drawing on Sun's study, the $Ni(OH)_2@Cu(OH)_2$-CF electrode was fabricated through a single-step solvothermal process [1]. As is typical, 40 mL methanol, containing 240 mg $Ni(NO_3)_2·6H_2O$, was transferred into a 100 mL polytetrafluoroethylene autoclave. The as-synthesized $Cu(OH)_2$-CF was placed into methanol solution. Meanwhile, the steel autoclave was maintained at 180 °C for 4 h and allowed to cool down naturally in atmospheric conditions. Then, the reacted foam was rinsed with DI water three times and dried in air. The obtained foam was named as $Ni(OH)_2@Cu(OH)_2$-CF. Finally, the obtained precursor, $Ni(OH)_2@Cu(OH)_2$-CF, was annealed at 350 °C with a heating rate of 5 °C/min for 2 h. After cooling down to room temperature, a black product was obtained, designated as $NiO@Cu_2O$-CF. To prove the synthesis of NiO, we duplicated the methanol solution preparation and conducted the reaction under standard conditions, excluding the $Cu(OH)_2$-CF. Subsequently, the produced powder was subjected to calcination at 350 °C for a duration of 2 h to yield the NiO powder.

2.3. Structural Characterization

The crystal structure of sensitive materials was characterized by X-ray diffraction (XRD) on a MiniFlex 600 X-ray diffractometer (600 W, Rigaku, Tokyo, Japan). The morphology of $Ni(OH)_2@Cu(OH)_2$-CF and $NiO@Cu_2O$-CF were characterized by scanning an electron microscope (VEGA3, TESCAN, Bmo, Czech Republic). The hierarchical structure and chemical compositions of $NiO@Cu_2O$-CF were analyzed by transmission electron microscopy (TEM) and energy dispersive spectroscopy (EDS), respectively. The distribution of the elements of $NiO@Cu_2O$-CF was obtained by using the X-ray photoelectron spectroscopy (XPS, Thermo) spectrum. The specific surface area of $Cu(OH)_2$-CF and $NiO@Cu_2O$-CF were measured using the Tristar 3020 (Micromeritics instrument (Shanghai) Ltd., Norcross, GA, USA) physical adsorption instrument from Micromeritics.

2.4. Electrochemical Measurement

All the electrochemical experiments were completed using the electrochemical work station (HY-E1002A, Shenzhen Haoyang Technology Co., Ltd., Shenzhen, China). The Ag/AgCl (3 M KCl) electrode served as the reference, while the platinum electrode functioned as the counter electrode. The electrodes that were synthesized were employed as the working electrodes. Moreover, the electrochemical performances of obtained electrodes for glucose detection were evaluated by cyclic voltammetry and chronoamperometry tests.

During the test cycle voltammetry, the scan rate was set at 50 mV/s, with the voltage sweep spanning 0 to 0.9 V.

3. Results and Discussion

3.1. Morphological and Structural Characterization

The SEM analysis was conducted to examine the morphologies of CF, $Cu(OH)_2$-CF, $Ni(OH)_2$@$Cu(OH)_2$-CF and NiO@Cu_2O-CF. The typical three-dimensional structure of CF can be observed in Figure 1a. Figure 1b,c depict the structure of $Cu(OH)_2$-CF, featuring a uniform array of $Cu(OH)_2$ nanowires that have colonized the copper foam's surface. Clearly seen in Figure 1d–f, the $Cu(OH)_2$ nanowires are enveloped by a hydrangea-like layer. The formed $Ni(OH)_2$@$Cu(OH)_2$-CF composite features a rod-like hierarchical architecture and a distinct hydrangea-like appearance. For the production of metal oxide-based sensory materials, the $Ni(OH)_2$@$Cu(OH)_2$-CF was subjected to air annealing, leading to the oxidation of nickel and copper hydroxides to NiO and Cu_2O, respectively. As depicted in Figure 1g–i, the annealed NiO@Cu_2O-CF maintained its morphology with minimal alteration, resembling its pre-annealed state.

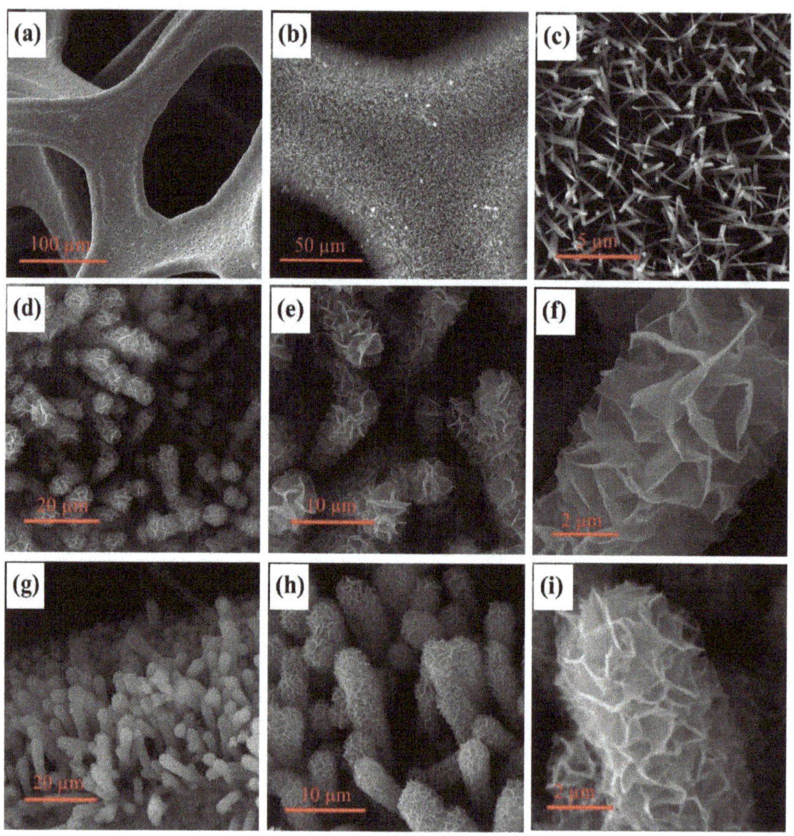

Figure 1. SEM image of CF (**a**), $Cu(OH)_2$ nanowires (**b**,**c**), $Ni(OH)_2$@$Cu(OH)_2$-CF (**d**–**f**) and NiO@Cu_2O-CF (**g**–**i**).

Further visualization of the morphology and structural attributes of $NiO@Cu_2O$-CF was achieved through TEM analysis and element mapping. The fragmented $NiO@Cu_2O$, which was scraped from the surface of the CF substrate, was characterized by TEM as shown in Figure 2a. Clearly, the nanowire (diameter $\approx$ 350 nm) is surrounded by an ultrathin shell featuring a folded structure. The evident interface between the nanowire and shells indicates a typical hierarchical core–shell structure. Figure 2b displays the high-resolution TEM result of the core–shell structural $NiO@Cu_2O$. The lattice space of 0.25 nm can correspond to the (111) plane of Cu_2O, and the 0.20 nm was indexed as the (200) plane of NiO. In addition, the selected area electron diffraction (SAED) result in Figure 2c revealed that the direction rings can correspond to (200), (220) planes of NiO and (111) planes of Cu_2O. Moreover, Figure 2d–f demonstrate the distribution of the Cu, Ni and O elements on $NiO@Cu_2O$.

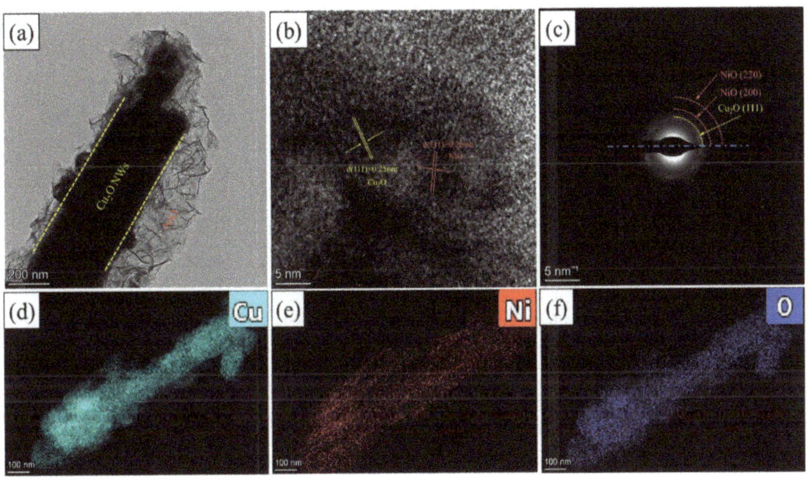

Figure 2. The TEM (**a**), HRTEM (**b**) and SAED (**c**) of NiO/Cu_2O; The elemental mapping of Cu (**d**), Ni (**e**) and O (**f**).

Furthermore, the XPS technique was employed to analyze the chemical composition of $NiO/Cu_2O/CF$. The complete spectrum in Figure 3a demonstrates the existence of Cu, Ni and O. The Cu 2p spectra in Figure 3b presents the fitted peaks with the binding energy of 931.6 eV, and 933.4 eV can be indexed as the Cu 2p3/2, evidencing the existence of Cu^+. And the peaks of 952.5 eV are classified as the Cu 2p1/2. Figure 3c reveals the Ni 2p spectral region with Ni 2p 3/2 peaks located at 853 eV and 854.7 eV, while the peak at 871.8 eV is attributed to Ni 2p1/2, indicative of Ni^{2+} ions.

To further study the crystallographic structure of as-fabricated samples, the X-ray powder diffraction (XRD) pattern delineation of $Cu(OH)_2$-CF, $Ni(OH)_2@Cu(OH)_2$-CF, $NiO@Cu_2O$-CF and NiO powder were carried out and are shown in Figure 3d. It was found that the XRD pattern of as-synthesized materials agreed well with the standard PDF card of Cu (PDF#04-0836), $Cu(OH)_2$ (PDF#013-0420), $Ni(OH)_2$ (PDF#014-0117), Cu_2O (PDF#05-0667) and NiO (PDF#47-1049), respectively. For $NiO@Cu_2O$-CF, the NiO peak is faint. This may be due to the small amount of coverage on the nanowires. Consequently, an identical method was employed to synthesize the NiO powder. There are three peaks located at 36.5°, 42.4° and 61.6° which correspond to the (111), (200) and (220) crystal planes of Cu_2O [41,42]. The diffraction peaks at 37.2°, 43.2°, 62.8°, 75.4° and 79.4° can be indexed to the (111), (200) (220), (311) and (222) crystal planes, respectively, of NiO [43]. The XRD data confirm the effective synthesis of the materials.

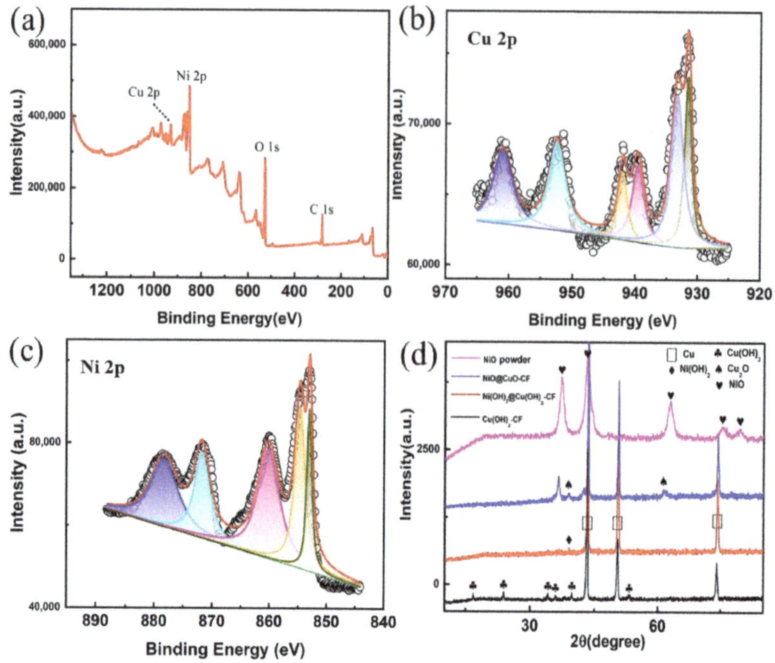

Figure 3. The chemical composition and crystalline structure analysis of NiO@Cu_2O-CF: survey spectrum of XPS (**a**), Cu 2p spectrum of XPS (**b**), and Ni 2p spectrum of XPS (**c**); (**d**) the XRD patterns of as-synthesized materials.

3.2. Electrochemical Properties of NiO@Cu_2O-CF

To facilitate testing, the synthesized sensitive material was trimmed into a 0.5 × 1.5 cm rectangle and subsequently secured with electrode clips for subsequent electrochemical experiments. And Figure 4a depicts the measurement setup of the working electrode. The electrochemical behavior of the synthesized electrodes was assessed using cyclic voltammetry. Figure 4b illustrates the CV results of as-fabricated electrodes, which were tested in 0.1 M NaOH with 5.0 mM and without glucose. For $Cu(OH)_2$-CF, the oxidation current shows a minimal rise in the presence and absence of glucose. As the $Cu(OH)_2$ nanowires were coated with $Ni(OH)_2$ nanosheets, there was an enhancement in the electrocatalytic efficiency towards glucose, though the oxidation current observed a minimal rise, signifying that the $Ni(OH)_2$ nanosheets have a limited catalytic capacity for glucose oxidation. When the as-fabricated $Ni(OH)_2$@$Cu(OH)_2$-CF was oxidized to NiO@Cu_2O-CF, the CV exhibited obvious current increase once the glucose was added to the solution. Comparative experiments reveal that NiO/Cu_2O exhibits superior electrochemical catalytic activity towards glucose compared to $Cu(OH)_2$ and $Ni(OH)_2$@$Cu(OH)_2$. Additionally, as depicted in Figure 4c, there is a marked rise in the electrocatalytic current in response to increasing glucose concentrations, confirming that the NiO@Cu_2O-CF electrode is sensitive to different glucose levels and is suitable for use as a glucose sensor.

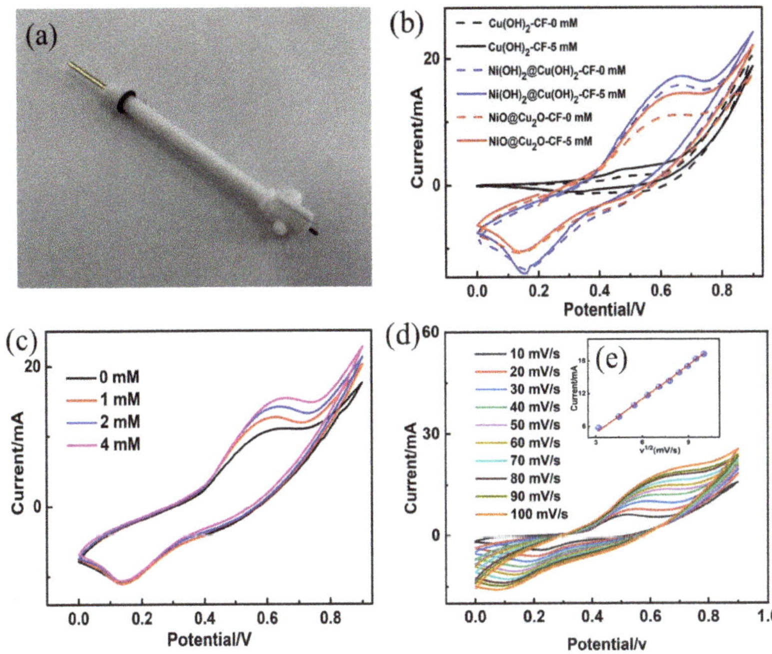

Figure 4. (**a**) The measurement setup of the working electrode; (**b**) CV curves (vs. Ag/AgCl) of $Cu(OH)_2$-CF, $Ni(OH)_2$@$Cu(OH)_2$-CF and NiO@Cu_2O-CF with 0 and 5.0 mM glucose in 0.1 M NaOH at 50 mV/s; (**c**) CVs (vs. Ag/AgCl) of NiO/Cu_2O/CF with 0, 2.0 and 6.0 mM glucose; (**d**) the CV results that dependent on different scan rate (10–100 mV/s) in 0.1 M NaOH with 5.0 mM glucose; (**e**) inset image is the linear fitting of catalytic current increment vs. the square root of scan rate ($\nu^{1/2}$).

The mechanism of electrochemically catalyzed glucose on NiO@Cu_2O-CF is accompanied by the formation of CuOOH and NiOOH in NaOH solution and reacts with hydroxyl anions to form glucolactone [3,37,44]. In other words, the Ni^{2+} and Cu^{+} of NiO and Cu_2O are oxidized to Ni^{3+} and Cu^{3+} in test solution. With the glucose added, glucose on the surface of NiO@Cu_2O was oxidated via deprotonation. Meanwhile, the Ni^{3+} and Cu^{3+} was reduced to Ni^{2+} and Cu^{2+}; the reactions can be expressed as follows [45]:

$$NiO + OH^- \rightarrow NiOOH + e^- \quad (1)$$

$$Cu_2O + 2OH^- + H_2O \rightarrow 2Cu(OH)_2 + 2e^- \quad (2)$$

$$Cu(OH)_2 + OH^- \rightarrow CuOOH + H_2O + e^- \quad (3)$$

$$NiOOH + glucose \rightarrow NiO + gluconolactone \quad (4)$$

$$CuOOH + glucose \rightarrow Cu(OH)_2 + gluconolactone \quad (5)$$

Furthermore, the electrochemical catalytic kinetics of NiO/Cu_2O/CF were analyzed in 0.1 mM NaOH solution with 5.0 mM glucose under different scan rates (10–100 mV/s). As Figure 4c shows, the oxidation peak current increases gradually and presents a positive correlation with the scanning rate. Figure 4d,e displays a liner relationship between the anodic peak current and the square root of scan rates ($\nu^{1/2}$), which suggest the electrochemical reaction is a typical diffusion-controlled process.

3.3. Potentiostatic Catalytic Properties of $NiO@Cu_2O$-CF

The amperometry technique was used to further estimate the electrochemical sensing characteristics of $NiO@Cu_2O$-CF. Figure 5a shows the amperometric responses of $NiO@Cu_2O$-CF at potential variations from 0.40 to 0.55 V in 0.1 M NaOH solution. When the potential was 0.40 V, the current response after adding glucose was very small and no obvious step shape was observed. Moreover, the background current will increase significantly as the applied potential increases. At the same times, it is considered that the higher the potential, the more interferences will be electrocatalyzed, resulting in a decrease in the selectivity. Therefore, 0.45 V is determined as the optimal potential. Subsequently, the typical i–t curve in Figure 5b was obtained by successively dropping the glucose solution to 50 mL 0.1 M NaOH with a 20 s interval at 0.45 V. According to the results, the $NiO@Cu_2O$-CF presents a larger measurement scale for the glucose range from 0.005 mM to 4.5 mM. The inset image shows a low glucose concentration for the amperometric titration test range of 5 μM to 60 μM, which demonstrated that the electrode responded sensitively, even at a low glucose concentration. The quadratic polynomial fitting curve of catalytic current versus glucose concentration was defined as $y\ (\text{mA}) = 0.00118x\ (\mu\text{M}) - 0.00008x^2(\mu\text{M}) + 0.000033$ ($R^2 = 0.9985$) in Figure 5c. Additionally, Figure 5d displays the linear fitting curve between current and low glucose concentration with the equation: $y\ (\text{mA}) = 0.00116x\ (\mu\text{M}) + 0.000034$ ($R^2 = 0.9861$). According to the fitting linear and geometric area of the electrode, the as-fabricated $NiO@Cu_2O$-CF-based sensor possesses a sensitivity of 4.67 $\mu A/\mu M/cm^2$.

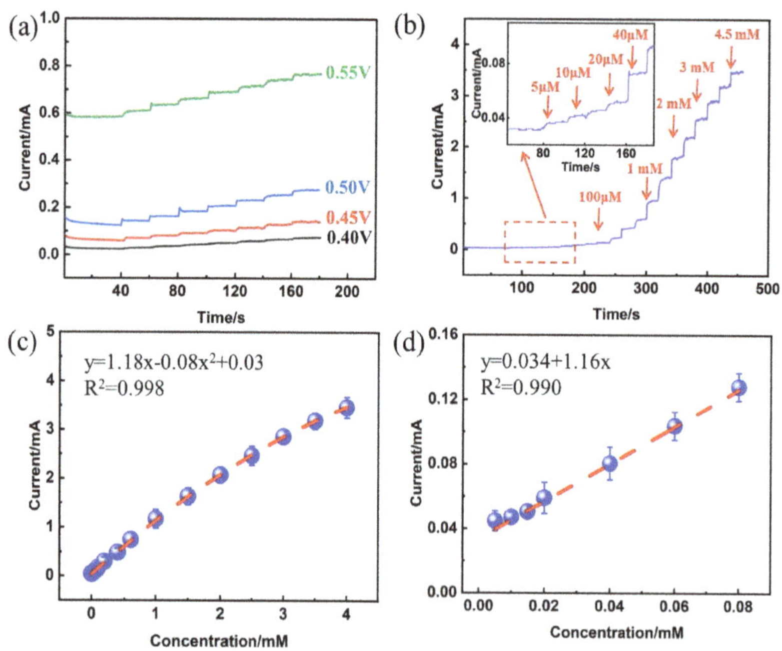

Figure 5. (**a**) Amperometric test curves of $NiO@Cu_2O$-CF with continuous adding of glucose under different potentials: 0.40 V, 0.45 V, 0.50 V and 0.55 V; (**b**) amperometric titration experiment of $NiO@Cu_2O$-CF at 0.45 V with 0.005–4.5 mM glucose, and the inset is the i–t curve in low glucose concentrations; (**c**) the quadratic polynomial fitting calibration line of the sensor; (**d**) the linear fitting calibration line of the sensor.

Additionally, as well as sensitivity, the response–recovery behavior, anti-interference performance, reproducibility and stability also make a great difference in evaluating the electrochemical catalytic ability for an efficient sensor. Figure 6a recorded the response–recovery ability of NiO@Cu_2O-CF when the glucose was increased and diluted. At first, the current increased sharply when the glucose solution was dropped into the electrochemical cell. Then, the glucose was diluted by adding NaOH solution and the catalytic current decreased. From the inset image of Figure 6a, it took about 1.0 s to reach a steady-state value in the reaction cell, and it recovered to the base line in about 3.5 s. This confirms that the NiO@Cu_2O-CF possesses fast response and recovery characteristics in sensing glucose. In addition, the effect of the deformed electrode was detected by bending the electrode from 0° to 120°. According to the result shown in Figure 6b, there was no significant difference for the CV in the NaOH solution with 4 mM glucose, suggesting the good anti-bending performance of the as-constructed electrode.

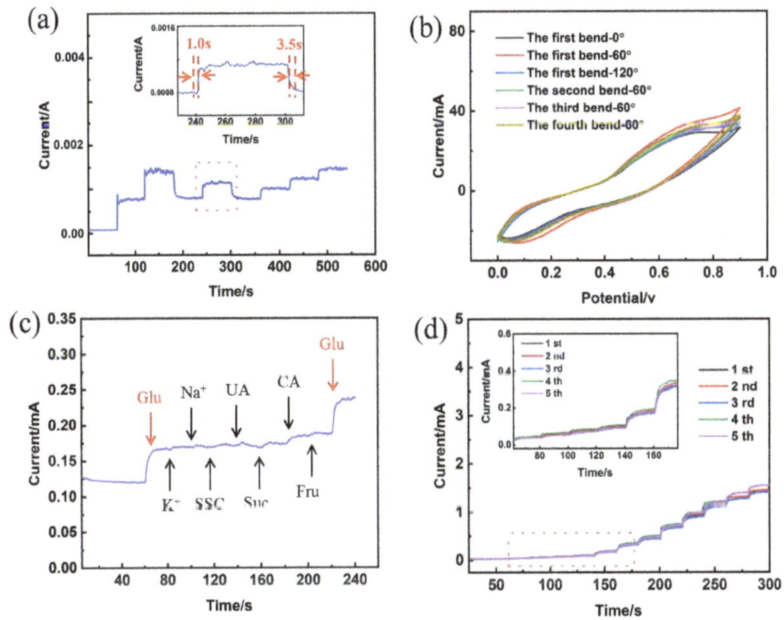

Figure 6. (**a**) The i–t test curve of NiO@Cu_2O-CF when increasing and diluting the glucose concentration, and the inset image shows the response and recovery times toward 0.75 mM glucose; (**b**) the CVs of NiO@Cu_2O-CF when it was bent to different angles (0°, 60° and 120°). (**c**) Interference test of NiO@Cu_2O-CF when adding KCl, NaCl, sodium citrate, urea, sucrose, citric acid and fructose; (**d**) the recorded i–t test results of five sensors prepared by the same process and inset is the response current in low concentration.

Selectivity is an essential ability for NiO@Cu_2O-CF in practical application as an electrochemical sensor. As shown in Figure 6c, the interferences, such as KCl, NaCl, sodium citrate, urea, sucrose, citric acid and fructose, were added to the electrochemical cell in sequence. The concentration of the added interference was 100 μM. Compared to adding glucose, there was no significant current response with injecting those substances into the electrochemical cell. This indicates that the NiO@Cu_2O-CF electrode possesses excellent selectivity for glucose determination. To evaluate the reproducibility of NiO@Cu_2O-CF, we carried out an amperometry experiment on five electrodes fabricated by the same process. As shown by Figure 6d, almost all the electrodes respond with a similar current change with the glucose concentration range being 0.04–3.0 mM. The inset image is the results at a low glucose concentration range (0.04–0.4 mM). For evaluating the long-term stability of NiO@Cu_2O-CF, cyclic voltammetry test was carried out on the same electrode in an electrochemical cell containing 5 mM glucose over 30 days (Figure 7). The relative standard deviation (RSD) of the current was calculated as 1.5% with 30 days, which demonstrated that the as-synthesized sensor presented outstanding long-term stability.

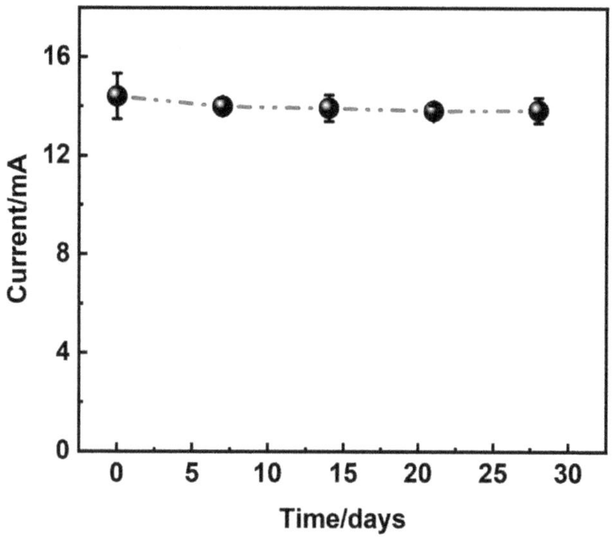

Figure 7. Long-term stability of NiO@Cu_2O-CF electrode.

Finally, we compared this work with previous works which were developed using nickel oxide and copper oxide nanomaterials. We listed the key parameters in Table 1. We can conclude that the fabricated NiO@Cu_2O-CF electrode exhibits good performance for glucose detection, including the high sensitivity and large linear range. Through the BET test, we found that the specific surface area of NiO@Cu_2O-CF (4.26 m^2/g) increased significantly compared with that of $Cu(OH)_2$-CF (2.43 m^2/g). Therefore, the possible reason is that the hierarchical core–shell structure can effectively improve the electrochemical performance for oxide nanomaterials.

Table 1. A performance summary of nickel oxide and copper oxide-based nonenzymatic glucose sensors.

Materials	Sensitivity ($\mu A/\mu M/cm^2$)	Linear Range (μM)	Ref.
N-CuO/Cu_2O:NiO film	1.13	0.05–2.74	[46]
g-C_3N_4/NiO/CuO	0.362	0.0004–8.5	[47]
NiO/CuO	112.08 μA/mM	0.14–180.00 mM	[48]
CuO/NiO@GCE	0.0397	0.04–4.86	[49]
CuO/NiO-0.4NFs	1.32	0.01–10.0	[50]
Ni-CuO/Cu/Cu_2O	0.208	0.01–8.0	[51]
NiO@Cu_2O-CF	4.67	0.005–4.5	This work

4. Conclusions

In summary, we successfully fabricated a hierarchical core–shell structure named NiO@Cu_2O-CF, in which $Cu(OH)_2$ nanowires was encapsulated by $Ni(OH)_2$ nanosheets on the Cu foam substrate. Then, we smartly treated $Ni(OH)_2$@$Cu(OH)_2$-CF at 350 °C to obtain NiO@Cu_2O-CF with the same framework. The electrochemical detections demonstrated that the NiO@Cu_2O-CF electrode presented the property of excellent glucose sensing, including high sensitivity (1.49 $\mu A/\mu M/cm^2$), a large linear range (0.005–4.5 mM) and fast response/recovery times (1.0 s/3.5 s). In addition, the NiO@Cu_2O-CF electrode also exhibited good reproducibility, selectivity, anti-folding performance and long-term stability. Therefore, the as-synthesized electrode has great potential and prospects to work as an enzyme-free glucose sensor.

Author Contributions: Conceptualization, Y.H. and Z.Z.; methodology, Y.H., J.Y. and Y.D.; validation, J.Y. and Y.D.; formal analysis, Y.H., J.Y. and Y.D.; investigation, Y.H. and Z.Z.; resources, Y.X., T.W., F.Z. and W.G.; data curation, Y.X., T.W., F.Z. and W.G.; writing—original draft preparation, Y.H. and Z.Z.; writing—review and editing, Z.Z.; supervision, Y.H. and Z.Z.; funding acquisition, Z.Z. All authors have read and agreed to the published version of the manuscript.

Funding: This research was funded by the National College Student Innovation and Entrepreneurship Project (No. 202310577016) and the Key Project of Universities in Guangdong Province (No. 2022ZDZX3031).

Data Availability Statement: Data are contained within the article.

Conflicts of Interest: The authors declare no conflicts of interest.

References

1. Sun, S.; Shi, N.; Liao, X.; Zhang, B.; Yin, G.; Huang, Z.; Chen, X.; Pu, X. Facile synthesis of CuO/$Ni(OH)_2$ on carbon cloth for non-enzymatic glucose sensing. *Appl. Surf. Sci.* **2020**, *529*, 147067. [CrossRef]
2. Mao, W.; He, H.; Sun, P.; Ye, Z.; Huang, J. Three-Dimensional Porous Nickel Frameworks Anchored with Cross-Linked Ni(OH)(2) Nanosheets as a Highly Sensitive Nonenzymatic Glucose Sensor. *ACS Appl. Mater. Interfaces* **2018**, *10*, 15088–15095. [CrossRef] [PubMed]
3. Manikandan, A.; Veeramani, V.; Chen, S.-M.; Madhu, R.; Lee, L.; Medina, H.; Chen, C.-W.; Hung, W.H.; Wang, Z.M.; Shen, G.; et al. Low-Temperature Chemical Synthesis of Three-Dimensional Hierarchical $Ni(OH)_2$-Coated Ni Microflowers for High Performance Enzyme-Free Glucose Sensor. *J. Phys. Chem. C* **2016**, *120*, 25752–25759. [CrossRef]
4. Lu, P.; Liu, Q.; Xiong, Y.; Wang, Q.; Lei, Y.; Lu, S.; Lu, L.; Yao, L. Nanosheets-assembled hierarchical microstructured $Ni(OH)_2$ hollow spheres for highly sensitive enzyme-free glucose sensors. *Electrochim. Acta* **2015**, *168*, 148–156. [CrossRef]
5. Lu, P.; Lei, Y.; Lu, S.; Wang, Q.; Liu, Q. Three-dimensional roselike alpha Ni(OH)(2) assembled from nanosheet building blocks for non-enzymatic glucose detection. *Anal. Chim. Acta* **2015**, *880*, 42–51. [CrossRef]
6. Caldara, M.; Lowdon, J.W.; van Wissen, G.; Ferrari, A.G.M.; Crapnell, R.D.; Cleij, T.J.; Diliën, H.; Banks, C.E.; Eersels, K.; van Grinsven, B. Dipstick Sensor Based on Molecularly Imprinted Polymer-Coated Screen-Printed Electrodes for the Single-Shot Detection of Glucose in Urine Samples—From Fundamental Study toward Point-of-Care Application. *Adv. Mater. Interfaces* **2023**, *10*, 2300182. [CrossRef]
7. Adeel, M.; Rahman, M.M.; Caligiuri, I.; Canzonieri, V.; Rizzolio, F.; Daniele, S. Recent advances of electrochemical and optical enzyme-free glucose sensors operating at physiological conditions. *Biosens. Bioelectron.* **2020**, *165*, 112331. [CrossRef]
8. Cui, Y.; Duan, W.; Jin, Y.; Wo, F.; Xi, F.; Wu, J. Ratiometric Fluorescent Nanohybrid for Noninvasive and Visual Monitoring of Sweat Glucose. *ACS Sens.* **2020**, *5*, 2096–2105. [CrossRef]
9. Du, P.; Niu, Q.; Chen, J.; Chen, Y.; Zhao, J.; Lu, X. "Switch-On" Fluorescence Detection of Glucose with High Specificity and Sensitivity Based on Silver Nanoparticles Supported on Porphyrin Metal–Organic Frameworks. *Anal. Chem.* **2020**, *92*, 7980–7986. [CrossRef] [PubMed]
10. Wen, J.; Li, N.; Li, D.; Zhang, M.; Lin, Y.; Liu, Z.; Lin, X.; Shui, L. Cesium-Doped Graphene Quantum Dots as Ratiometric Fluorescence Sensors for Blood Glucose Detection. *ACS Appl. Nano Mater.* **2021**, *4*, 8437–8446. [CrossRef]
11. Lin, C.; Du, Y.; Wang, S.; Wang, L.; Song, Y. Glucose oxidase@Cu-hemin metal-organic framework for colorimetric analysis of glucose. *Mater. Sci. Eng. C* **2021**, *118*, 111511. [CrossRef]

12. Tong, L.; Wu, L.; Zai, Y.; Zhang, Y.; Su, E.; Gu, N. Paper-based colorimetric glucose sensor using Prussian blue nanoparticles as mimic peroxidase. *Biosens. Bioelectron.* **2023**, *219*, 114787. [CrossRef] [PubMed]
13. Gao, Y.; Huang, Y.; Chen, J.; Liu, Y.; Xu, Y.; Ning, X. A Novel Luminescent "Nanochip" as a Tandem Catalytic System for Chemiluminescent Detection of Sweat Glucose. *Anal. Chem.* **2021**, *93*, 10593–10600. [CrossRef]
14. Yang, H.; Liu, J.; Feng, X.; Nie, F.; Yang, G. A novel copper-based metal-organic framework as a peroxidase-mimicking enzyme and its glucose chemiluminescence sensing application. *Anal. Bioanal. Chem.* **2021**, *413*, 4407–4416. [CrossRef]
15. Li, D.; Tan, R.; Mi, X.; Fang, C.; Tu, Y. An electrochemiluminescent biosensor for noninvasive glucose detection based on cluster-like AuAg hollowed-nanoparticles. *Microchem. J.* **2021**, *167*, 106271. [CrossRef]
16. Saha, T.; Del Caño, R.; Mahato, K.; De la Paz, E.; Chen, C.; Ding, S.; Yin, L.; Wang, J. Wearable Electrochemical Glucose Sensors in Diabetes Management: A Comprehensive Review. *Chem. Rev.* **2023**, *123*, 7854–7889. [CrossRef]
17. Sehit, E.; Altintas, Z. Significance of nanomaterials in electrochemical glucose sensors: An updated review (2016–2020). *Biosens. Bioelectron.* **2020**, *159*, 112165. [CrossRef] [PubMed]
18. Teymourian, H.; Barfidokht, A.; Wang, J. Electrochemical glucose sensors in diabetes management: An updated review (2010–2020). *Chem. Soc. Rev.* **2020**, *49*, 7671–7709. [CrossRef]
19. Govindaraj, M.; Srivastava, A.; Muthukumaran, M.K.; Tsai, P.C.; Lin, Y.C.; Raja, B.K.; Rajendran, J.; Ponnusamy, V.K.; Arockia Selvi, J. Current advancements and prospects of enzymatic and non-enzymatic electrochemical glucose sensors. *Int. J. Biol. Macromol.* **2023**, *253 Pt 2*, 126680. [CrossRef]
20. Nguyen, T.N.H.; Jin, X.; Nolan, J.K.; Xu, J.; Le, K.V.H.; Lam, S.; Wang, Y.; Alam, M.A.; Lee, H. Printable Nonenzymatic Glucose Biosensors Using Carbon Nanotube-PtNP Nanocomposites Modified with AuRu for Improved Selectivity. *ACS Biomater. Sci. Eng.* **2020**, *6*, 5315–5325. [CrossRef]
21. Lin, L.; Weng, S.; Zheng, Y.; Liu, X.; Ying, S.; Chen, F.; You, D. Bimetallic PtAu alloy nanomaterials for nonenzymatic selective glucose sensing at low potential. *Electroanal. Chem.* **2020**, *865*, 114147. [CrossRef]
22. Wang, R.; Liu, X.; Zhao, Y.; Qin, J.; Xu, H.; Dong, L.; Gao, S.; Zhong, L. Novel electrochemical non-enzymatic glucose sensor based on 3D Au@Pt core–shell nanoparticles decorated graphene oxide/multi-walled carbon nanotubes composite. *Microchem. J.* **2022**, *174*, 107061. [CrossRef]
23. Zhao, Z.; Sun, Y.; Huang, Y.; Jin, J.; Wei, X.; Gong, W.; Hu, J.; Chen, Y. In situ construction of zeolitic imidazolate framework-67 derived Co_3O_4 on CCCH NWs/CF hierarchical nanowires for high-performance enzymeless glucose detection. *Microchem. J.* **2021**, *160*, 105623. [CrossRef]
24. Zhang, D.; Shao, Y.; Kong, X.; Jiang, M.; Lei, D.; Lei, X. Facile fabrication of large-area hybrid Ni-Co hydroxide/$Cu(OH)_2$/copper foam composites. *Electrochim. Acta* **2016**, *218*, 294–302. [CrossRef]
25. Zhao, Z.; Huang, Y.; Li, Q.; Mei, H.; Zhu, F.; Gong, W. Electrochemical glucose sensitive device based on graphene supported Co_3O_4@Ag NWs core-shell nanostructures. *Appl. Surf. Sci.* **2021**, *565*, 150553. [CrossRef]
26. Cao, W.; Guo, T.; Ding, Y.; Hu, Y.; Liu, D. Substantially boosting the performance of non-enzymatic glucose sensing: Introducing a novel CuS nanosheet anchored on hollow $Ni(OH)_2$ nanosphere. *Appl. Surf. Sci.* **2023**, *634*, 157650. [CrossRef]
27. Cheng, D.; Wang, T.; Zhang, G.; Wu, H.; Mei, H. A novel nonenzymatic electrochemical sensor based on double-shelled $CuCo_2O_4$ hollow microspheres for glucose and H_2O_2. *J. Alloy. Compd.* **2020**, *819*, 153014. [CrossRef]
28. Xiao, X.; Zhang, Z.; Nan, F.; Zhao, Y.; Wang, P.; He, F.; Wang, Y. Mesoporous $CuCo_2O_4$ rods modified glassy carbon electrode as a novel non-enzymatic amperometric electrochemical sensors with high-sensitive ascorbic acid recognition. *J. Alloy. Compd.* **2021**, *852*, 157045. [CrossRef]
29. Babulal, S.M.; Chen, S.-M.; Palani, R.; Venkatesh, K.; Haidyrah, A.S.; Ramaraj, S.K.; Yang, C.-C.; Karuppiah, C. Graphene oxide template based synthesis of $NiCo_2O_4$ nanosheets for high performance non-enzymatic glucose sensor. *Colloid. Surface A* **2021**, *621*, 126600. [CrossRef]
30. Jang, K.-b.; Park, K.R.; Kim, K.M.; Hyun, S.-k.; Ahn, C.; Kim, J.C.; Lim, S.-c.; Han, H.; Mhin, S. Electrochemical performance of the spinel $NiCo_2O_4$ based nanostructure synthesized by chemical bath method for glucose detection. *Appl. Surf. Sci.* **2021**, *545*, 148927. [CrossRef]
31. Imanzadeh, H.; Amiri, M.; Nozari-Asbemarz, M. A novel NiO/C@ rGO nanocomposite derived from Ni (gallate): A non-enzymatic electrochemical glucose sensor. *Microchem. J.* **2024**, *199*, 110106. [CrossRef]
32. Thu, D.T.P.; Xuan, C.T.; Oanh, V.T.; Viet, N.X.; Tu, L.M.; Loan, T.T.; Van Tuan, C.; Ha, N.H.; Le, D.T.T.; Hoa, N.D. One-Step Method for the Direct Growth of NiO Nanoflowers on Pencil Graphite Electrode for Highly Sensitive Non-Enzymatic Glucose Sensing. *J. Electrochem. Soc.* **2024**, *171*, 017505.
33. de Luna, G.S.; Tabanelli, T.; Velasco-Vélez, J.J.; Monti, E.; Ospitali, F.; Albonetti, S.; Cavani, F.; Fornasari, G.; Benito, P. Electrification of glucose valorization over NiO/Ni foam. *Sustain. Energ. Fuels* **2023**, *7*, 4474–4485. [CrossRef]
34. Jiang, L.-C.; Zhang, W.-D. A highly sensitive nonenzymatic glucose sensor based on CuO nanoparticles-modified carbon nanotube electrode. *Biosens. Bioelectron.* **2010**, *25*, 1402–1407. [CrossRef]

35. Pu, F.; Miao, H.; Lu, W.; Zhang, X.; Yang, Z.; Kong, C. High-performance non-enzymatic glucose sensor based on flower-like Cu_2O-Cu-Au ternary nanocomposites. *Appl. Surf. Sci.* **2022**, *581*, 152389. [CrossRef]
36. Waqas, M.; Wu, L.; Tang, H.; Liu, C.; Fan, Y.; Jiang, Z.; Wang, X.; Zhong, J.; Chen, W. Cu_2O Microspheres Supported on Sulfur-Doped Carbon Nanotubes for Glucose Sensing. *ACS Appl. Nano Mater.* **2020**, *3*, 4788–4798. [CrossRef]
37. Yang, H.; Bao, J.; Qi, Y.; Zhao, J.; Hu, Y.; Wu, W.; Wu, X.; Zhong, D.; Huo, D.; Hou, C. A disposable and sensitive non-enzymatic glucose sensor based on 3D graphene/Cu_2O modified carbon paper electrode. *Anal. Chim. Acta* **2020**, *1135*, 12–19. [CrossRef] [PubMed]
38. Zhou, X.; Guo, W.; Yao, Y.; Peng, R.; Peng, P. Flexible Nonenzymatic Glucose Sensing with One-Step Laser-Fabricated Cu_2O/Cu Porous Structure. *Adv. Eng. Mater.* **2021**, *23*, 2100192. [CrossRef]
39. Li, M.; Meng, J.; Li, Q.; Huang, M.; Liu, X.; Owusu, K.A.; Liu, Z.; Mai, L. Finely Crafted 3D Electrodes for Dendrite-Free and High-Performance Flexible Fiber-Shaped Zn–Co Batteries. *Adv. Funct. Mater.* **2018**, *28*, 1802016. [CrossRef]
40. Xu, H.; Shi, Z.X.; Tong, Y.X.; Li, G.R. Porous microrod arrays constructed by carbon-confined NiCo@ $NiCoO_2$ core@ shell nanoparticles as efficient electrocatalysts for oxygen evolution. *Adv. Mater.* **2018**, *30*, 1705442. [CrossRef] [PubMed]
41. Wang, L.; Duan, C.; Miao, X.; Liu, J.; Qu, Y.; Gao, J.; Wang, B.; Yin, Z. Free-Standing and High-Sensitive Electrodes with Hierarchical Nanostructures of Bimetallic Hydroxides M(OH)x/$Cu(OH)_2$/CF (M=Ni, Co, Fe and Zn) for Glucose Detection. *ChemistrySelect* **2021**, *6*, 3576–3583. [CrossRef]
42. Pinsker, Z.G.; Imamov, R.M. Production and investigation of thin films of Cu_2O. *Sov. Phys.-Crystallogr.* **1964**, *9*, 334–335.
43. Nakahigashi, K.; Fukuoka, N.; Shimomura, Y. Crystal Structure of Antif erromagnetic NiO Determined by X-Ray Topography. *J. Phys. Soc. Jpn.* **1975**, *38*, 1634–1640. [CrossRef]
44. Huang, W.; Ge, L.; Chen, Y.; Lai, X.; Peng, J.; Tu, J.; Cao, Y.; Li, X. $Ni(OH)_2$/NiO nanosheet with opulent active sites for high-performance glucose biosensor. *Sens. Actuat. B-Chem.* **2017**, *248*, 169–177. [CrossRef]
45. Li, G.; Wang, X.; Liu, L.; Liu, R.; Shen, F.; Cui, Z.; Chen, W.; Zhang, T. Controllable synthesis of 3D $Ni(OH)_2$ and NiO nanowalls on various substrates for high-performance nanosensors. *Small* **2015**, *11*, 731–739. [CrossRef] [PubMed]
46. Palmer, M.; Masikini, M.; Jiang, L.; Wang, J.; Franscious, C.; Chamier, J.; Inyang, O.; Chowdhury, M. Enhanced electrochemical glucose sensing performance of CuO:NiO mixed oxides thin film by plasma assisted nitrogen doping. *J. Alloy. Compd.* **2021**, *853*, 156900. [CrossRef]
47. Lotfi, Z.; Gholivand, M.B.; Shamsipur, M. Non-enzymatic glucose sensor based on a g-C_3N_4/NiO/CuO nanocomposite. *J. Anal. Biochem.* **2021**, *616*, 114062. [CrossRef] [PubMed]
48. Razmi, H.; Safikhanlou, H.; Maleki, F. Electrosynthesis of tunable NiO/CuO-printed circuit board electrodes for glucose sensing. *Microchem. J.* **2024**, *206*, 111460. [CrossRef]
49. Muthulakshmi, L.; Annaraj, J.; Chang, P.; Selvaraj, M.; Singh, G.; Arumugam, B. Bioflocculant polymer reduced CuO/NiO binary transition metal oxide nanocomposite: Application as an effective non-enzymatic glucose sensor. *Inorg. Chem. Commun.* **2024**, *170*, 113250. [CrossRef]
50. Li, M.; Huan, K.; Deng, D.; Yan, X.; Li, Y.; Luo, L. Coaxial electrospinning synthesis of size tunable CuO/NiO hollow heterostructured nanofibers: Towards detection of glucose level in human serum. *Colloids Surf. B* **2023**, *222*, 113047. [CrossRef]
51. Uzhandaivel, H.; Paramasivam, K.; Manickam, S.; Nallathambi, K.S. Nickel-doped CuO/Cu/Cu_2O nanocomposite as an efficient electrode for electrochemical non-enzymatic glucose sensor and asymmetric supercapacitor. *J. Appl. Electrochem.* **2023**, *53*, 1869–1886. [CrossRef]

nanomaterials

MDPI

Article

Effect of Bias Voltage on the Microstructure and Photoelectric Properties of W-Doped ZnO Films

Haijuan Mei [1,*], Wanli Wang [2], Junfeng Zhao [1,*], Weilong Zhong [1], Muyi Qiu [1], Jiayang Xu [1], Kailin Gao [1], Ge Liu [1], Jianchu Liang [1] and Weiping Gong [1,*]

[1] Guangdong Provincial Key Laboratory of Electronic Functional Materials and Devices, Huizhou University, Huizhou 516007, China

[2] School of Civil Engineering and Architecture, Nanchang Jiaotong University, Nanchang 330100, China

* Correspondence: haijuanmei@hzu.edu.cn (H.M.); zhaojunfeng@hzu.edu.cn (J.Z.); gwp@hzu.edu.cn (W.G.)

Abstract: W-doped ZnO (WZO) films were deposited on glass substrates by using RF magnetron sputtering at different substrate bias voltages, and the relationships between microstructure and optical and electrical properties were investigated. The results revealed that the deposition rate of WZO films first decreased from 8.8 to 7.1 nm/min, and then increased to 11.5 nm/min with the increase in bias voltage. After applying a bias voltage to the substrate, the bombardment effect of sputtered ions was enhanced, and the films transformed from a smooth surface into a compact and rough surface. All the films exhibited a hexagonal wurtzite structure with a strong (002) preferred orientation and grew along the c-axis direction. When the bias voltage increased, both the residual stress and lattice parameter of the films gradually increased, and the maximum grain size of 43.4 nm was achieved at −100 V. When the bias voltage was below −300 V, all the films exhibited a high average transmittance of ~90% in the visible light region. As the bias voltage increased, the sheet resistance and resistivity of the films initially decreased and then gradually increased. The highest F_{OM} of $5.8 \times 10^{-4}\ \Omega^{-1}$ was achieved at −100 V, possessing the best comprehensive photoelectric properties.

Keywords: WZO; bias voltage; microstructure; photoelectric properties

1. Introduction

Transparent conductive oxide (TCO) films often have a high conductive performance, high transmittance in the visible light wavelength range, and high reflectivity in the near-infrared light wavelength range. Due to its superior photoelectric properties, the TCO film has a broad application value in the field of photoelectric industry [1–3]. As the most widely used TCO film, tin-doped indium oxide (ITO) film has the advantages of high conductivity and high transmittance [4–6], but the raw material indium is scarce, expensive, and toxic. Reducing or avoiding the consumption of indium has become the inevitable development trend of transparent conductive film. With the development of optoelectronic technology, higher requirements are put forward for the performance of transparent electrode sin flexible displays. As a wide-band-gap semiconductor metal oxide, ZnO film has the advantages of cheap raw materials, rich reserves, and being non-toxic and easy to bend, meaning it is a promising TCO film material. However, the carrier concentration of the undoped ZnO film is not high and the conductivity is poor, meaning appropriate doping is needed to improve the conductivity of the films [7,8].

In recent years, by doping some high-valence elements into ZnO films, more electrons and holes have been obtained to form the N-type semiconductor, thus increasing the carrier concentration and the conductivity of the films, such as Al-doped ZnO (AZO) [9], Ga-doped ZnO (GZO) [10], F-doped ZnO (FZO) [11], Mo-doped ZnO (MZO) [12], and W-doped ZnO (WZO) [13] films. Due to the high valence difference between W^{6+} and

Zn^{2+}, each W atom can contribute four extra electrons to the electrical conductivity. These additional electrons increase the carrier concentration, thereby reducing the resistivity of the ZnO film and making it more conductive. Moreover, W^{6+} (0.062 nm) has a similar ionic radius to Zn^{2+} (0.074 nm), and it is easy to replace Zn atoms with W without causing significant lattice distortion [14]. Thus, the electrical properties of ZnO films can be effectively improved by doping them with W, as has been widely applied in various fields, including transparent conductive oxides, solar cells, and UV detectors. For example, Abliz et al. [15] deposited WZO films by using radiofrequency (RF) magnetron sputtering, and they found that the appropriate doping of W efficiently reduced the oxygen vacancy defects, carrier concentration, and total trap density, which resulted in good electrical properties and high stability. In addition to element doping, the thickness of WZO films also has a significant impact on its photoelectric properties; for instance, the lowest resistivity together with a high transmittance of 90% were achieved at 332 nm [16]. During the sputtering deposition, process parameters such as substrate temperature [17] and bias voltage [18–21] have a significant impact on the structure and properties of the films. As an important deposition parameter, the substrate bias voltage is often used to accelerate the sputtered particles to obtain higher energy, promote the migration of atoms on the substrate surface, and help to form a dense microstructure, so as to improve the quality of the films. However, there are few reports on the effect of substrate bias voltage on the photoelectric properties of WZO films.

In this study, W-doped ZnO films were deposited by RF magnetron sputtering at different substrate bias voltages, and the relationships between microstructure and optical and electrical properties were explored.

2. Experimental Details

2.1. Coating Deposition

WZO films were deposited on ordinary glasses by RF magnetron sputtering using a ceramic target of WZO (99.99% purity, ZnO:W = 98:2 wt%). A rotating substrate holder was placed at the center of the chamber with a rotation speed of 5 rpm, and the distance between the target and the substrate was 115 mm. All the substrates were ultrasonically cleaned in acetone and alcohol for 20 min, respectively, and then placed on a rotating substrate holder subsequent to desiccation. Prior to the deposition, the chamber was pumped to 8.0×10^{-4} Pa, and then heated up to 200 °C. High-purity Ar gas was introduced into the chamber, and the flow rate and working pressure were fixed at 80 sccm and 0.5 Pa, respectively. Then, the WZO films were prepared by RF magnetron sputtering for 60 min. During the deposition process, the target power was controlled at 150 W. As the only variable, the substrate bias voltage was varied from 0 to −300 V, and the detailed deposition parameters are shown in Table 1.

Table 1. Deposition parameters of WZO films.

Parameters	
Base pressure (Pa)	8.0×10^{-4}
Deposition temperature (°C)	200
Working pressure (Pa)	0.5
Ar flow rate (sccm)	80
Target-to-substrate distance (mm)	115
WZO target	ZnO:W = 98:2 wt%
Target power (W)	150
Substrate bias voltage (V)	0, −100, −200, −300
Deposition time (min)	60

2.2. Coating Characterization

The surface and cross-section morphologies, as well as thickness of the films were characterized by scanning electron microscopy (SEM, Tescan Vega 3 Xmu, Brno, Czech Republic). The crystalline structure of the films was characterized by X-ray diffraction (XRD, Bruker D8 advance, Karlsruhe, Germany) using Cu $K\alpha$ radiation (λ = 0.15406 nm) in the $\theta/2\theta$ mode, where the scanning angle and scanning step were 30°–60° and 0.01°, respectively. Based on biaxial strain model analysis [22], Bragg's law [23], and the Scherrer equation [24], the residual stresses (σ), lattice parameters (c), and grain sizes (D) of films along the c-axis of the (002) plane can be determined, respectively.

$$\sigma = -233 \times \frac{c - c_0}{c_0} \quad (1)$$

$$c = \frac{\lambda}{\sin\theta} \quad (2)$$

$$D = \frac{0.9\lambda}{\beta\cos\theta} \quad (3)$$

where λ, β, and c_0 = 5.2066 Å refer to the wavelength of the Cu $K\alpha$ radiation, the full width at half maximum (FWHM), and the unstrained lattice constant of ZnO along the c-axis, respectively. The optical transmittance spectrum from 300 to 1000 nm was measured by using a visible spectrophotometer (723PCSR, Ruifeng, Guangzhou, China). The sheet resistance and resistivity of the films were determined by a four-probe resistivity tester (FT-316B, Ruipin Instrument, Ningbo, China).

3. Results

3.1. Microstructure

Figure 1 presents surface micrographs of WZO films deposited at various bias voltages. As can be seen in Figure 1a, without applying bias voltage, the WZO film showed a compact and smooth surface. After applying a bias voltage of −100 V to the substrate, the surface became rough, accompanied by some large particles, as shown in Figure 1b. When the bias voltage was increased to −200 V, the WZO film showed a compact and rough surface morphology, which was mainly caused by ion bombardment and etching effects under high bias voltage. When the bias voltage was further increased to −300 V, the film presented a rod-like nanocrystal morphology on the surface, which corresponded to the top of the columnar crystal structure of the film. A similar surface structure was also found for the W-doped ZnO films deposited at high substrate temperatures [17]. After applying a bias voltage, the cations in the plasma will continuously bombard the substrate and the growth surface of the film under the action of the electric field, promoting the migration and aggregation of surface atoms and forming large grains. As the bias voltage increased, the ion bombardment effect was further enhanced, causing damage to the growth surface of the films and forming a compact and rough structure.

Figure 2 shows the cross-sectional micrographs of WZO films deposited at various bias voltages. As can be seen in Figure 2a, without applying bias voltage, the WZO film showed a fine columnar crystal structure, and the film thickness reached 527 nm. There are no obvious gaps at the interface between the film and the glass substrate, indicating good bonding between the film and the substrate. After applying a bias voltage of −100 V, the WZO film transformed into a dense columnar crystal structure, and the thickness decreased to 428 nm. This can be related to the enhanced ion bombardment effect after applying bias voltage, resulting in the formation of a thin and dense columnar crystal structure. However, when the bias voltage was further increased to −200 and −300 V, the thickness of the films gradually increased, and the films exhibited a distinct columnar crystal structure. As shown in Figure 3, when increasing the bias voltage, the deposition rate of WZO films first decreased from 8.8 to 7.1 nm/min, and then gradually increased to 11.5 nm/min at −300 V. After applying bias voltage, the ion bombardment effect was enhanced, the films became more compact, and this led to an initial decrease in the deposition rate. A similar result was also found for the TiN coatings deposited by magnetron sputtering at various bias voltages [25]. Similarly, the deposition rate of AlTiVCuN coatings initially decreased due to the enhanced ion bombardment effect at high ion source currents [26]. When the substrate bias voltage was further increased, more ions with high energy flew towards the growing surface of the film per unit of time, resulting in an increase in the deposition rate of the films.

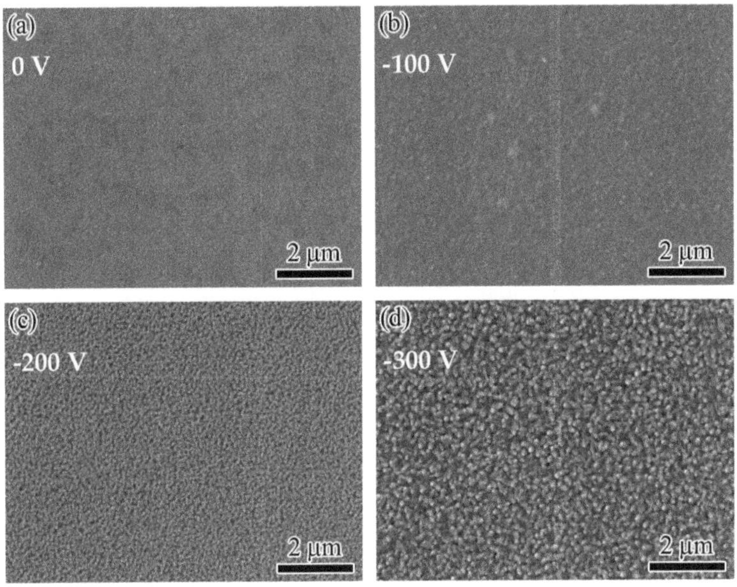

Figure 1. Surface micrographs of the WZO films at various bias voltages: (**a**) 0 V, (**b**) −100 V, (**c**) −200 V, (**d**) −300 V.

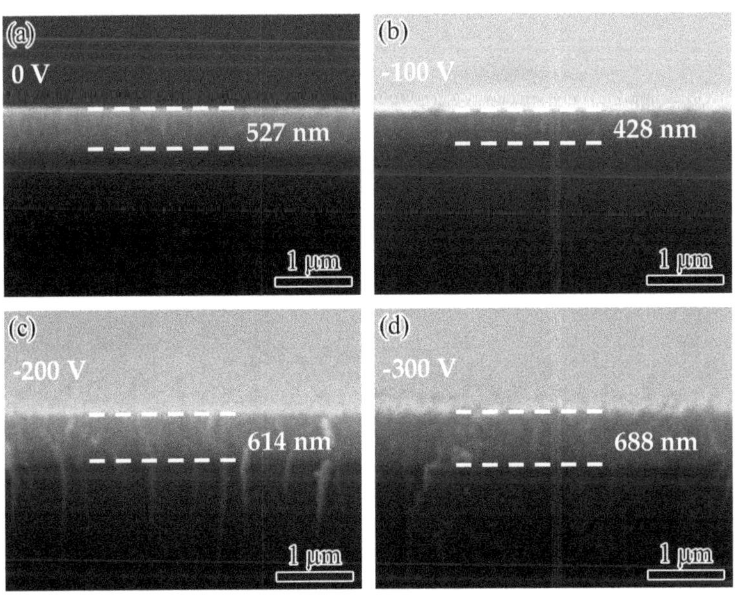

Figure 2. Cross-sectional micrographs of the WZO films at various bias voltages: (**a**) 0 V, (**b**) −100 V, (**c**) −200 V, (**d**) −300 V.

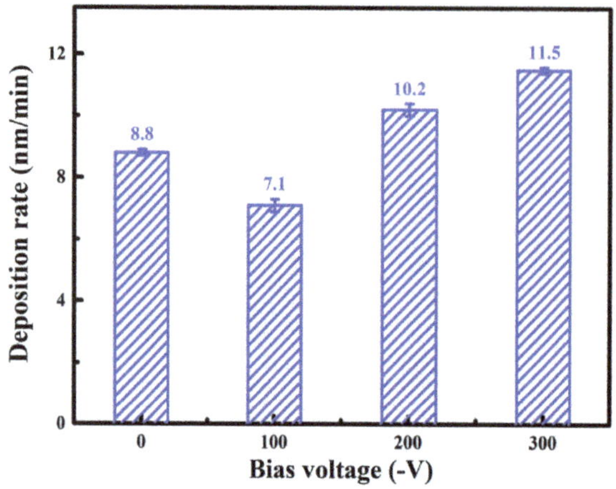

Figure 3. Deposition rates of the WZO films at various bias voltages.

Figure 4a displays the XRD pattern of WZO films deposited at various bias voltages. Without applying bias voltage, a strong diffraction peak appeared at about 34.35°, which corresponded to the (002) plane of the ZnO phase. This indicated that the WZO films showed a hexagonal wurtzite structure with a strong preferred orientation of the (002) plane, that they and grew along the c-axis direction. In addition, a weak diffraction peak appeared at about 35.97°, corresponding to the (101) plane of the ZnO phase, indicating that the WZO films exhibited a polycrystalline structure. However, no W-related phase appeared in the XRD pattern. It was demonstrated by XPS analysis that the W atoms were completely oxidized, and that they existed in the oxidized state of W^{6+} [17]. Due to the similar ionic radii of W^{6+} and Zn^{2+}, the doped W tended to replace Zn sites in the hexagonal wurtzite structure [14,15]. When compared to the standard diffraction peak of the ZnO phase (JCPDS 36-1451), both the (002) and (101) diffraction peaks of the films shifted towards lower angles. Generally, due to the difference in ionic radius, replacing Zn^{2+} (0.074 nm) with W^{6+} (0.062 nm) will cause a decrease in the lattice parameters, resulting in a shift of diffraction peaks towards higher angles. Thus, the leftward shift of diffraction peaks was mainly caused by the residual stresses in the films, including thermal stress and intrinsic stress [27]. With the increase in bias voltage, the (002) diffraction peak of WZO films gradually shifted towards lower angles, which was mainly related to the enhanced ion bombardment under higher bias voltages. The calculation of residual stresses and lattice parameters of the films was conducted by Equations (1) and (2), respectively (Figure 4b). All the films exhibited compressive residual stress, and the WZO films presented the lowest compressive residual stress of 0.48 GPa at 0 V. With increasing bias voltage, the residual stress gradually increased to 2.47 GPa at −300 V. Correspondingly, the lattice parameter also gradually increased from 5.2172 to 5.2618 Å. A similar phenomenon was found for the AZO/Cu/AZO films, where the lattice parameter showed a positive correlation with the residual compressive stress [23]. When increasing the bias voltage, the weak (101) peak disappeared, and the intensity of the (002) peak gradually decreased, indicating a decrease in crystallinity of WZO films, which could be due to the increase in lattice defects generated under the strong ion bombardment [28]. In addition, the full width at half-maximum (FWHM) of the (002) peak also varied with the bias voltage. Based on the Scherrer Equation (3), the grain size of the WZO films can be calculated, as shown in Figure 4c. As the bias voltage increased from 0 to −300 V, the grain size first increased from 27.1 to 43.4 nm, then gradually decreased to 24.1 nm, suggesting that the maximum grain size was achieved for the film deposited at −100 V. Under an appropriate bias voltage, the sputtered species with high kinetic energy continuously flew towards the substrate surface, which increased the mobility of surface atoms and then promoted atomic nucleation and growth.

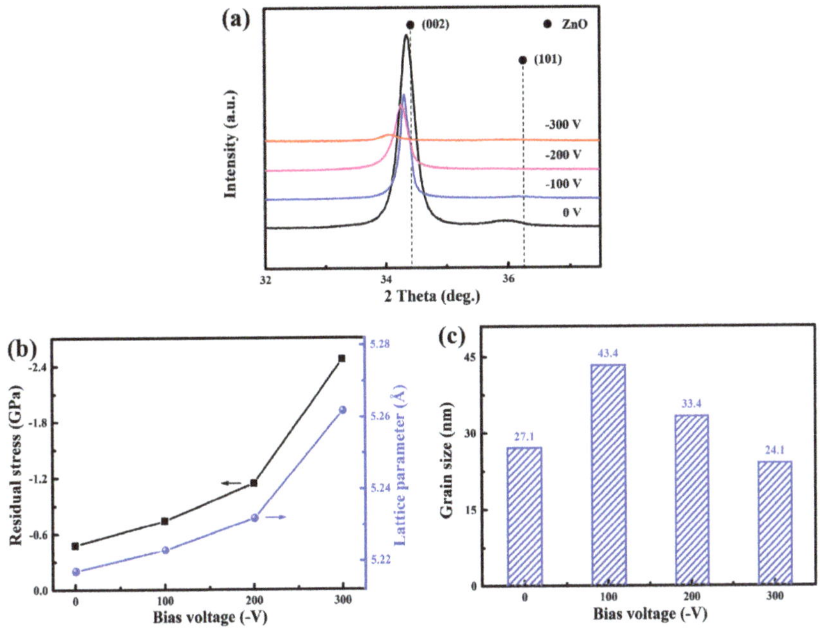

Figure 4. XRD patterns (**a**), residual stresses and lattice parameters (**b**), and grain sizes (**c**) of the WZO films at various bias voltages.

3.2. Photoelectric Properties

Figure 5a displays the transmission spectra of WZO films deposited at various bias voltages. It can be seen that all the transmission spectra exhibited obvious fluctuations, which can be related to the interference effect caused by the reflection effect at interfaces [16]. Similar fluctuating transmission spectra have also been observed in other W-doped ZnO films [17,29], and the optical band gap of WZO films ranged from 3.45 to 3.52 eV [30]. All the films have a sharp absorption edge in the ultraviolet range of 350–400 nm, and they shifted to a longer wavelength with the increase in bias voltage. Based on the transmittance in the range of 380–780 nm, the average transmittance of the films in the visible light region can be calculated, as shown in Figure 5b. At low bias voltages, all the WZO films exhibited a high average transmittance of ~90% in the visible light region. After applying a high bias voltage of −300 V, the average transmittance of WZO films decreased to 88.5%. This decrease in the average transmittance can be explained by the following reasons. Firstly, with the increase in bias voltage, the plasma bombardment effect was enhanced, resulting in an increase in the surface roughness of the films, which enhanced the scattering of photons, thereby reducing the optical transmittance. Secondly, the film thickness was increased at high bias voltages, which enhanced the absorption of photons, resulting in a decrease in average transmittance [17]. Thirdly, the crystallinity of the film decreased and the grain size became smaller at high bias voltages, resulting in an increase in defects and grain boundaries in the films, which enhanced the scattering effect of photons and thereby reduced the optical transmittance [31].

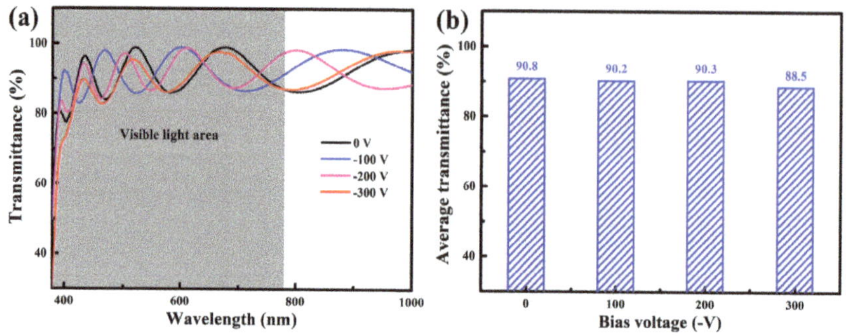

Figure 5. Transmittance (**a**) and average transmittance (**b**) of the WZO films at various bias voltages.

Figure 6 shows the sheet resistance and resistivity of the WZO films deposited at various bias voltages. As the bias voltage increased from 0 to −300 V, the sheet resistance of the films initially decreased from 1.1 to 0.6 kΩ/sq, and then gradually increased to 0.9 kΩ/sq. Correspondingly, the electrical resistivity of the films exhibited a similar trend of change with the increase in bias voltage. Thus, applying an appropriate bias voltage was conductive to improving the electrical conductivity of WZO films, and the lowest electrical resistivity of 2.6×10^{-2} Ω·cm was achieved at −100 V, which was lower than that of WZO films with different W doping concentrations [15]. The initially decrease in the sheet resistance and resistivity was mainly attributed to the increase in grain size, as confirmed by the XRD results mentioned above. The larger the grain size, the lower the density of grain boundaries, which reduced the grain boundary scattering and allowed electrons to migrate more freely in the crystal, thereby reducing the electrical resistivity of the films. A similar phenomenon was also found in WZO films deposited at various substrate temperatures, where the electrical resistivity decreased with the increase in grain size [17]. However, when the bias voltage increases above −100 V, the increase in electrical resistivity can be attributed to the poor crystallinity and the smaller grain size. Thin films with low crystallinity usually have more defects, which can capture or scatter free electrons, as well as restrict electron migration, thereby leading to an increase in the electrical resistivity. Similar results were also found in the W-doped ZnO [16] and Zr-doped ZnO [32] films, where the electrical conductivity increased with improved crystallinity.

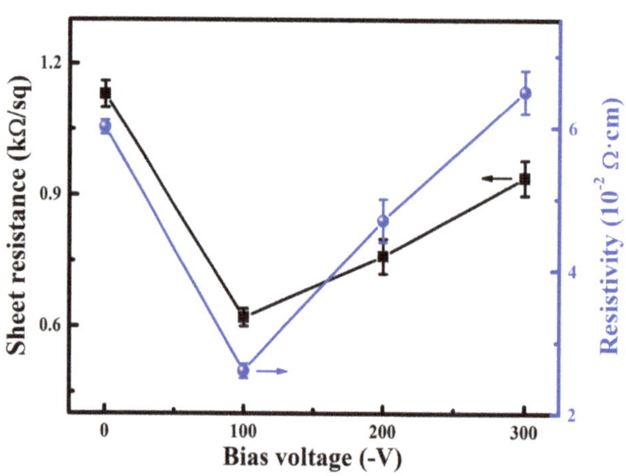

Figure 6. Sheet resistance and resistivity of the WZO films at various bias voltages.

For transparent conductive films, the figure of merit (F_{OM}) can be used to estimate the comprehensive photoelectric properties, which is defined as follows [33]:

$$F_{OM} = \frac{T_{av}^{10}}{R_S} \tag{4}$$

where T_{av} and R_s refer to the average optical transmittance in the visible region and the sheet resistance, respectively. Figure 7 presents the F_{OM} of WZO films deposited at various bias voltages. When increasing the bias voltage, the F_{OM} of WZO films exhibited a trend of first increasing and then decreasing. It can be clearly seen that the highest F_{OM} of $5.8 \times 10^{-4}\ \Omega^{-1}$ was achieved for the WZO film deposited at -100 V, which possessed a high transmittance of 90.2% as well as a low sheet resistance of 0.6 kΩ/sq. This indicated that the best comprehensive photoelectric properties of the films were achieved at a bias voltage of -100 V. As compared to the WZO films deposited at different substrate temperatures [17], the F_{OM} was relatively lower, which was mainly related to the high sheet resistance.

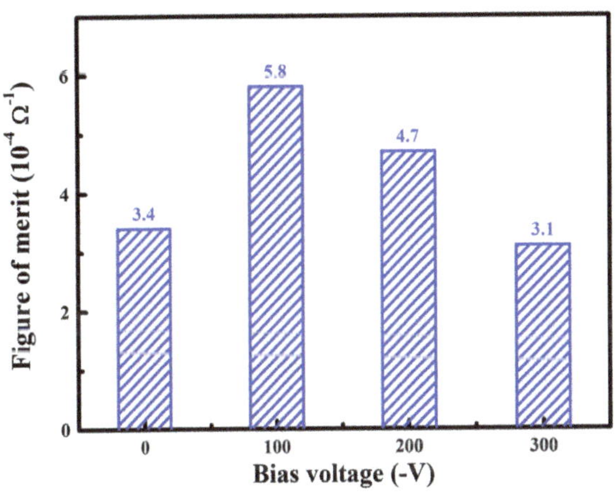

Figure 7. Figures of merit of the WZO films at various bias voltages.

4. Conclusions

In this study, W-doped ZnO films were deposited on glass substrates by RF magnetron sputtering. The bias voltage was found to have a significant impact on the structure and photoelectric properties of WZO films, providing a simple and effective process solution for preparing high-performance TCO films. The main conclusions are as follows:

(1) When the bias voltage increased from 0 to -300 V, the deposition rate of the films first decreased from 8.8 to 7.1 nm/min and then gradually increased to 11.5 nm/min. After applying bias voltage to the substrate, the ion bombardment effect was enhanced, and the surface morphology became much compact and rough.

(2) All the WZO films exhibited a hexagonal wurtzite structure with a strong (002) preferred orientation, and grew along the c-axis direction. With an increasing bias voltage, the lattice parameter gradually increased from 5.2172 to 5.2618 Å, and the grain size first increased from 27.1 to 43.4 nm and then gradually decreased to 24.1 nm.

(3) All the WZO films exhibited a high average transmittance of ~90% in the visible region. When the bias voltage was increased, the electrical resistivity of the films exhibited a trend of first decreasing and then increasing, and the highest F_{OM} of $5.8 \times 10^{-4}\ \Omega^{-1}$ was achieved at -100 V.

Author Contributions: Writing original manuscript, H.M. and W.W.; experiment, J.Z. and W.Z.; characterization, M.Q. and J.X.; data analysis, K.G. and G.L.; manuscript revision, J.L.; resources, W.G. All authors have read and agreed to the published version of the manuscript.

Funding: The present study was supported by the National Natural Science Foundation of China (52405189), the Guangdong-Hui Joint Fund Regional Cultivation Project (2023A1515140074), the Characteristic Innovation Project of Guangdong Provincial Universities (2024KTSCX087), the Key Construction Discipline Scientific Research Capacity Improvement Project in Guangdong Province (2021ZDJS079), Special Projects in Key Areas of Guangdong Provincial Department of Education (2023ZDZX1026), Key Research Projects of Huizhou University (HZU201901), the National College Student Innovation Training Program (202310577004), and the Innovative Research Team of Guangdong Province and Huizhou University (IRTHZU).

Data Availability Statement: Data are contained within the article.

Conflicts of Interest: The authors declare no conflicts of interest.

References

1. Lee, J.; Han, T.H.; Park, M.H.; Jung, D.Y.; Seo, J.; Seo, H.K.; Cho, H.; Kim, E.; Chung, J.; Choi, S.Y.; et al. Synergetic electrode architecture for efficient graphene-based flexible organic light-emitting diodes. *Nat. Commun.* **2016**, *7*, 11791. [CrossRef] [PubMed]
2. Kim, H.M.; Geng, D.; Kim, J.; Hwang, E.; Jang, J. Metal-oxide stacked electron transport layer for highly efficient inverted quantum-dot light emitting diodes. *ACS Appl. Mater. Interfaces* **2016**, *8*, 28727–28736. [CrossRef] [PubMed]
3. Tran, V.H.; Khan, R.; Lee, I.H.; Lee, S.H. Low-temperature solution-processed ionic liquid modified SnO2 as an excellent electron transport layer for inverted organic solar cells. *Sol. Energy Mater. Sol. Cells* **2018**, *179*, 260–269. [CrossRef]
4. Hotovy, J.; Hüpkes, J.; Böttler, W.; Marins, E.; Spiess, L.; Kups, T.; Smirnov, V.; Hotovy, I.; Kováč, J. Sputtered ITO for application in thin-film silicon solar cells: Relationship between structural and electrical properties. *Appl. Surf. Sci.* **2013**, *269*, 81–87. [CrossRef]
5. Bao, S.Y.; Deng, X.; Mao, F.; Zhong, N.; Yue, F.Y.; Sun, L.; Xiang, P.H.; Duan, C.G. Ultra-flat ITO films on mica for high temperature transparent flexible electrodes. *Ceram. Int.* **2020**, *46*, 2268–2272. [CrossRef]
6. Liu, S.; Xu, X.; Jiang, J. Flexible transparent ITO thin film with high conductivity and high-temperature resistance. *Ceram. Int.* **2024**, *50*, 47649–47654. [CrossRef]
7. Lung, C.; Toma, M.; Pop, M.; Marconi, D.; Pop, A. Characterization of the structural and optical properties of ZnO thin films doped with Ga, Al and (Al+Ga). *J. Alloys Compd.* **2017**, *725*, 1238–1243. [CrossRef]
8. Mallick, A.; Basak, D. Revisiting the electrical and optical transmission properties of co-doped ZnO thin films as n-type TCOs. *Prog. Mater. Sci.* **2018**, *96*, 86–110. [CrossRef]
9. Alev, O.; Özdemir, O.; Kılıç, A.; Akcan, D.; Büyükköse, S. Effect of Al doping on structural and optical properties of atomic layer deposited ZnO thin films. *Surf. Interfaces* **2024**, *52*, 104942. [CrossRef]
10. Lemine, A.S.; Bhadra, J.; Popelka, A.; Shakoor, R.A.; Ahmad, Z.; Al-Thani, N.J.; Hasan, A. Synergistic effect of concentration and annealing on structural, mechanical, and room-temperature thermoelectric properties of n-type Ga-doped ZnO films. *Ceram. Int.* **2024**, *50*, 47741–47753. [CrossRef]
11. Xing, L.; Lu, Y.; Liu, Y.; Wang, F.; Zhu, L.; Pan, X.; Ye, Z. Ga, Mg, F ternary-doped ZnO-based transparent conductive film prepared by radio-frequency sputtering with high transmittance. *Mater. Lett.* **2024**, *360*, 135972. [CrossRef]
12. Shkir, M.; Khan, M.T.; Khan, A. Impact of Mo doping on photo-sensing properties of ZnO thin films for advanced photodetection applications. *J. Alloys Compd.* **2024**, *985*, 174009. [CrossRef]
13. Ngom, B.D.; Mpahane, T.; Manyala, N.; Nemraoui, O.; Buttner, U.; Kana, J.B.; Fasasi, A.Y.; Maaza, M.; Beye, A.C. Structural and optical properties of nano-structured tungsten-doped ZnO thin films grown by pulsed laser deposition. *Appl. Surf. Sci.* **2009**, *255*, 4153–4158. [CrossRef]
14. Ngom, B.D.; Sakho, O.; Manyala, N.; Kana, J.B.; Mlungisi, N.; Guerbous, L.; Fasasi, A.Y.; Maaza, M.; Beye, A.C. Structural, morphological and photoluminescence properties of W-doped ZnO nanostructures. *Appl. Surf. Sci.* **2009**, *255*, 7314–7318. [CrossRef]
15. Abliz, A.; Wan, D.; Yang, L.; Mamat, M.; Chen, H.; Xu, L.; Wang, C.; Duan, H. Investigation on the electrical performances and stability of W-doped ZnO thin-film transistors. *Mat. Sci. Semicon. Proc.* **2019**, *95*, 54–58. [CrossRef]
16. Zhang, H.F.; Liu, H.F.; Lei, C.X.; Yuan, C.K.; Zhou, A.P. Tungsten-doped ZnO transparent conducting films deposited by direct current magnetron sputtering. *Vacuum* **2010**, *85*, 184–186. [CrossRef]
17. Zhang, C.; Chen, X.; Geng, X.; Tian, C.; Huang, Q.; Zhao, Y.; Zhang, X. Temperature-dependent growth and properties of W-doped ZnO thin films deposited by reactive magnetron sputtering. *Appl. Surf. Sci.* **2013**, *274*, 371–377. [CrossRef]
18. Elmkhah, H.; Zhang, T.F.; Abdollah-zadeh, A.; Kim, K.H.; Mahboubi, F. Surface characteristics for the Ti-Al-N coatings deposited by high power impulse magnetron sputtering technique at the different bias voltages. *J. Alloys Compd.* **2016**, *688*, 820–827. [CrossRef]
19. Ma, Q.; Li, L.; Xu, Y.; Gu, J.; Wang, L.; Xu, Y. Effect of bias voltage on TiAlSiN nanocomposite coatings deposited by HiPIMS. *Appl. Surf. Sci.* **2017**, *392*, 826–833. [CrossRef]

20. Sekhar, M.C.; Reddy, N.N.K.; Reddy, B.P.; Prakash, B.P.; Akkera, H.S.; Uthanna, S.; Park, S.H. Influence of substrate bias voltage on crystallographic structure, optical and electronic properties of Al/$(Ta_2O_5)_{0.85}(TiO_2)_{0.15}$/p-Si MIS Schottky barrier diodes fabricated by dc magnetron sputtering. *Mat. Sci. Semicon. Proc.* **2018**, *76*, 80–86. [CrossRef]
21. Jiang, X.; Zeng, X.K.; Xie, W.; Liu, M.; Leng, Y.X. Optimizing substrate bias voltage to improve mechanical and tribological properties of ductile FeCoNiCu high entropy alloy coatings with FCC structure. *J. Alloys Compd.* **2024**, *1004*, 175972. [CrossRef]
22. Tüzemen, E.S.; Eker, S.; Kavak, H.; Esen, R. Dependence of film thickness on the structural and optical properties of ZnO thin films. *Appl. Surf. Sci.* **2009**, *255*, 6195–6200. [CrossRef]
23. Mendil, D.; Challali, F.; Touam, T.; Bockelée, V.; Ouhenia, S.; Souici, A.; Djouadi, D.; Chelouche, A. Preparation of RF sputtered AZO/Cu/AZO multilayer films and the investigation of Cu thickness and substrate effects on their microstructural and optoelectronic properties. *J. Alloys Compd.* **2021**, *860*, 158470. [CrossRef]
24. Langford, J.I.; Wilson, A.J.C. Scherrer after sixty years: A survey and some new results in the determination of crystallite size. *J. Appl. Crystallogr.* **1978**, *11*, 102–113. [CrossRef]
25. Chun, S.Y. Bias voltage effect on the properties of TiN films by reactive magnetron sputtering. *J. Korean Phys. Soc.* **2010**, *56*, 1134–1139. [CrossRef]
26. Mei, H.; Yan, K.; Wang, R.; Cheng, L.; Li, Q.; Zhao, Z.; Ding, J.C.; Gong, W. Microstructure and mechanical properties of AlTiVCuN coatings prepared by ion source-assisted magnetron sputtering. *Nanomaterials* **2023**, *13*, 3146. [CrossRef]
27. Windischmann, H. Intrinsic stress in sputter-deposited thin-films. *Crit. Rev. Solid State* **1992**, *17*, 547–596. [CrossRef]
28. Wang, Y.P.; Lu, J.G.; Bie, X.; Ye, Z.Z.; Li, X.; Song, D.; Zhao, X.Y.; Ye, W.Y. Transparent conductive and near-infrared reflective Cu-based Al-doped ZnO multilayer films grown by magnetron sputtering at room temperature. *Appl. Surf. Sci.* **2011**, *257*, 5966–5971. [CrossRef]
29. Wang, Y.; Zhang, X.; Huang, Q.; Wei, C.; Zhao, Y. Room temperature deposition of highly conductive and transparent hydrogen and tungsten co-doped ZnO films for thin film solar cells applications. *Sol. Energy Mater. Sol. Cells* **2013**, *110*, 94–97. [CrossRef]
30. Zhang, H.; Yang, S.; Liu, H.; Yuan, C. Preparation and characterization of transparent conducting ZnO:W films by DC magnetron sputtering. *J. Semicond.* **2011**, *32*, 043002. [CrossRef]
31. Zhang, D.; Sun, K.; Yin, H.; Cheng, L.; Yuan, H.; Yang, C. Optimization of photoelectric properties and temporal stability of AZO/Ti/Cu/AZO films by insertion of Ti layer for low emissivity applications. *Mater. Sci. Eng. B* **2023**, *293*, 116471. [CrossRef]
32. Lv, M.; Xiu, X.; Pang, Z.; Dai, Y.; Han, S. Transparent conducting zirconium-doped zinc oxide films prepared by rf magnetron sputtering. *Appl. Surf. Sci.* **2005**, *252*, 2006–2011. [CrossRef]
33. Haacke, G. New figure of merit for transparent conductors. *J. Appl. Phys.* **1976**, *47*, 4086–4089. [CrossRef]

nanomaterials

Article

Research on the Microstructure, Mechanical Properties and Strengthening Mechanism of Nanocrystalline Al-Mo Alloy Films

Ying Wang [1], Huanqing Xu [1], Yulan Chen [1], Xiaoben Qi [1,*] and Ning Zhong [2]

[1] School of Materials Science, Shanghai Dianji University, Shanghai 201306, China; ywang@sdju.edu.cn (Y.W.); hh369792@126.com (H.X.); 15933618107@163.com (Y.C.)
[2] College of Ocean Science and Engineering, Shanghai Maritime University, Shanghai 201306, China; ningzhong@shmtu.edu.cn
* Correspondence: qixb@sdju.edu.cn

Abstract: In this work, the Al-Mo nanocrystalline alloy films with Mo contents ranging from 0–10.5 at.% were prepared via magnetron co-sputtering technology. The composition and microstructure of alloy thin films were studied using XRD, TEM, and EDS. The mechanical behaviors were tested through nanoindentation. The weights of each strengthening factor were calculated and the strengthening mechanism of alloy thin films was revealed. The results indicate that a portion of Mo atoms exist in the Al lattice, forming a solid solution of Mo in Al. The other part of Mo atoms tends to segregate at the grain boundaries, and this segregation becomes more pronounced with an increase in Mo content. There are no compounds or second phases present in any alloy films. As the Mo element content increases, the grain size of the alloy films gradually decreases. The hardness of pure aluminum film is 2.2 GPa. The hardness increases with an increase in Mo content. When the Mo content is 10.5 at.%, The hardness of the film increases to a maximum value of 4.9 GPa. The fine grain (ΔH_{gb}), solid solution (ΔH_{ss}), and nanocrystalline solute pinning ($\Delta H_{nc,ss}$) are the three main reasons for the increase in the hardness of alloy thin films. The contribution of ΔH_{gb} is the largest, accounting for over 60% of the total, while the contribution of ΔH_{ss} accounts for about 30%, ranking second. The rest of the increase is due to $\Delta H_{nc,ss}$.

Keywords: nanocrystalline; Al-Mo alloy film; microstructure; mechanical properties; strengthening mechanism

1. Introduction

Due to their excellent performance, nanocrystalline materials have been increasingly applied in fields such as aerospace and electronic information. Their application has always been a hot research topic in the academic community [1–4]. When the grain size decreases to the nanometer level, the volume fraction of the grain boundaries will increase exponentially. The stability of nanocrystalline material systems will be significantly reduced, making it easy for spontaneous grain growth to occur. For example, pure nanocrystalline metals such as Cu [5], Ni [6], and Al [7] will exhibit grain growth under external loading or at very low heating temperatures. Some nanocrystalline pure metal grains may even coarsen at room temperature [8]. The instability of their structure limits the application of nanocrystalline pure metals.

Therefore, improving the stability of nanocrystalline metals has always been an important research direction [9–11]. Some studies have shown that adding alloying elements to pure nanocrystalline metal is an effective way to improve their stability. Devaraj [12] studied the effect of the addition of Mg on the stability of nanocrystalline Al. It was found that, after annealing at 300 °C for 3 h, the grain size of the pure Al film doubled, while the size of the Al-10 at.% Mg alloy film only increased by 69%. Rajagopalan et al. [13] added Ta to nanocrystalline Cu. The grain size of the Cu-Ta nanocrystalline alloy did not show significant growth even after annealing at 400 °C. In addition, similar research results were reported for the Cu-Zr [14], FeNi-Zr [15], and Ni-W [16] systems. These additional alloying elements can limit the coarsening of these grains by reducing the system's free energy and pinning the nanocrystalline boundaries, thereby improving stability [17]. Therefore, compared to pure metals, nanocrystalline alloys have more practical value. It is also of greater research significance to study their mechanical behavior and corresponding reinforcement mechanism.

However, the strengthening mechanism of nanocrystalline alloys is also more complex [18–20]. With the change in alloy element content, the grain size, solute content, and properties of the grain boundaries of the nanocrystalline alloys will undergo corresponding changes. The interweaving of these microstructural factors makes it extremely difficult to reveal the strengthening mechanism [21]. Distinguishing the roles of various structural factors and studying them separately is an effective way to reveal the strengthening mechanism of nanocrystalline alloys. Some studies [20,22,23] have already attempted to do this through theoretical calculations. However, most of these studies obtained nanocrystalline alloys with fixed compositions and calculated the effects of various strengthening factors. Few systematic studies have been conducted on a series of nanocrystalline alloys to investigate the changing trends and weights of various strengthening factors during the alloy content variation process.

In this study, Al-Mo nanocrystalline alloy films were synthesized via magnetron sputtering technology. The effect of different molybdenum element contents on the microstructure and comprehensive mechanical properties of the alloy systems was studied in detail. Subsequently, the roles of various strengthening factors in the alloy films were calculated. The strengthening mechanism of nanocrystalline Al-Mo alloy films is revealed, based on the effects of various microstructural factors.

2. Materials and Methods

2.1. Film Deposition

The Al-Mo nanocrystalline alloy films with a Mo content ranging from 0 to 10.5 at.% were synthesized via multi-target magnetron co-sputtering technology using an ANAVA SPC-350 magnetron sputtering instrument (Anelva, Tokyo, Japan). The Al and Mo targets with a diameter of 75 mm were controlled by two RF cathodes, respectively. The distance between the target material and the substrate was 50 mm. The single-crystal Si wafers with dimensions of 15 mm × 15 mm were cleaned with alcohol ultrasonic for 15 min and dried before being placed on substrate racks in vacuum chambers. After evacuating the chamber to 5×10^{-4} Pa, the high-purity Ar gas, with a purity of 99.999%, was introduced. During the sputtering process, the working pressure was maintained at 6×10^{-1} Pa. A series of Al-Mo alloy thin films with a Mo content ranging from 0 to 10.5 at.% were prepared by changing the Mo target power while keeping the Al target power at 200 W. The Mo target powers were 0, 10, 20, 30, 40, and 50 W, respectively. The substrate was not heated or subjected to negative bias throughout the entire deposition process. The total deposition time was controlled at 2 h. The thickness of all alloy films was controlled at around 2 μm.

2.2. Films Characterization

The elemental contents (Al, Mo) of the alloy films were measured using an EDS (EDX, Oxford instruments INCA) (Oxford Instruments, Oxford, UK) attached to the scanning electron microscope (S-3400N SEM, Hitachi, Tokyo, Japan). The phase structure of the Al-Mo nanocrystalline alloy films was determined via X-ray diffraction (XRD) with Cu–Kα radiation using a D8 X-ray diffractometer (Bruker, Karlsruhe, Germany). The scanning range of 2θ was 30–70°, with a scanning speed of 5°/min. The microstructure was characterized using a JEOL-2100F TEM microscope (JEOL, Tokyo, Japan). The TEM samples were prepared using the soluble salt substrate method [24]. The comprehensive mechanical properties of the Al-Mo alloy systems were determined using a Step 300-NTH3 nano indentation (Anton Paar, Graz, Austria). The maximum load applied was 10 mN. The loading and unloading speed was 30 mN/min. The holding time was 10 s. The hardness and elastic modulus information of the films was analyzed using the Oliver–Pharr method [25]. To ensure the validity of the data, all samples were measured at more than 20 points.

3. Results and Discussions

3.1. The Microstructure of Alloy Films

Figure 1 presents the XRD pattern of the Al-Mo alloy films. As shown in the figure, all films contained a set of FCC Al peaks and no other Al compound peaks were found. This indicates that the film formed an Al-Mo solid solution. The XRD pattern of the pure Al film and the Al-0.9 at.% Mo film shows sharp diffraction peaks of Al (111) and Al (200), as well as weak diffraction peaks of Al (220). As the Mo content further increases, the Al (111) peak gradually widens while the Al (200) peak gradually weakens, indicating a gradual decrease in grain size. This is because the continuous addition of Mo atoms causes a large amount of lattice distortion in the Al film, which destroys the integrity of the Al grains. The diffraction peaks of Al (200) and Al (220) disappeared after the Mo element content increased to 10.5 at.% while the Al (111) peaks also became broadened and diffused. This indicates that the Al-Mo alloy film contains nanocrystalline or amorphous structures. Additionally, it is worth noting that as the Mo content increases, all diffraction peaks shifted towards larger angles. This is due to the substitution of larger-radius Al atoms with smaller-radius Mo atoms, resulting in the contraction of the Al lattice.

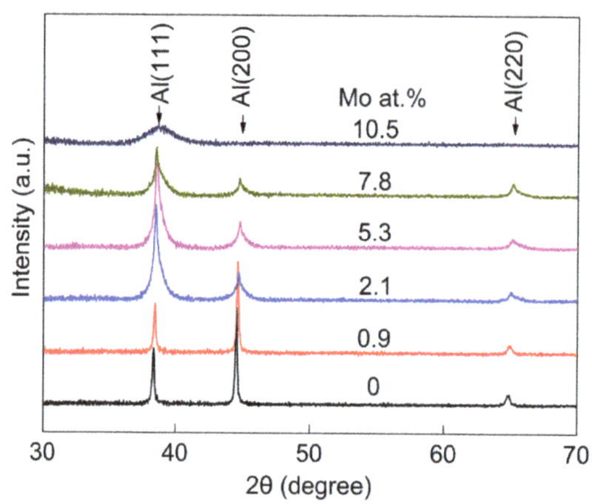

Figure 1. The XRD analysis results of Al-Mo alloy thin films.

According to the XRD results, the films' grain size and interplanar spacing were calculated using Scherrer formula and Bragg's law, respectively. In Figure 2a, the film grain size continues to decrease while the Mo content increases. In pure Al film, the grain size is about 121 nm. When the Mo element content increases to 0.9 at.%, it rapidly decreases to about 84 nm. Further increasing the Mo content leads to a continuous decrease in the grain size of the alloy film. When the Mo element content reaches 10.5 at.%, the grain size of the alloy film decreases to about 19 nm. This is because the high non-equilibrium of magnetron sputtering causes the sputtered particles to quickly lose kinetic energy when deposited on the substrate. The limitations of the dynamic conditions force Mo atoms to remain in the Al lattice. The addition of the Mo atoms causes severe distortion of the Al lattice, resulting in a decrease in grain size. In Figure 2b, the films' interplanar spacing continuously decreases with an increase in Mo contents. The dashed line represents the fitting curve.

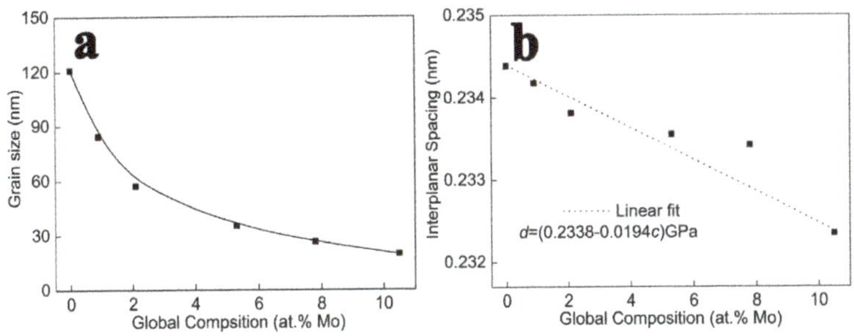

Figure 2. The relationship between grain size (**a**) and the interplanar spacing (**b**) of alloy thin films and Mo content.

In order to further determine the microstructure of the Al-Mo alloy films, a plane observation of the alloy films was performed using TEM. The results are shown in Figure 3. According to the grain size distribution diagrams in the lower right corner of Figure 3a–c, the grain size was about 120 nm in the pure Al film. The grain size continued to decrease with the increase in Mo. In the Al-7.8 at.% Mo alloy film, the grain size was only a dozen nanometers. In addition, the HRTEM analysis indicated that the interplanar spacing of the Al lattice gradually decreases with the increase in Mo content. This trend is consistent with the results of the XRD analysis.

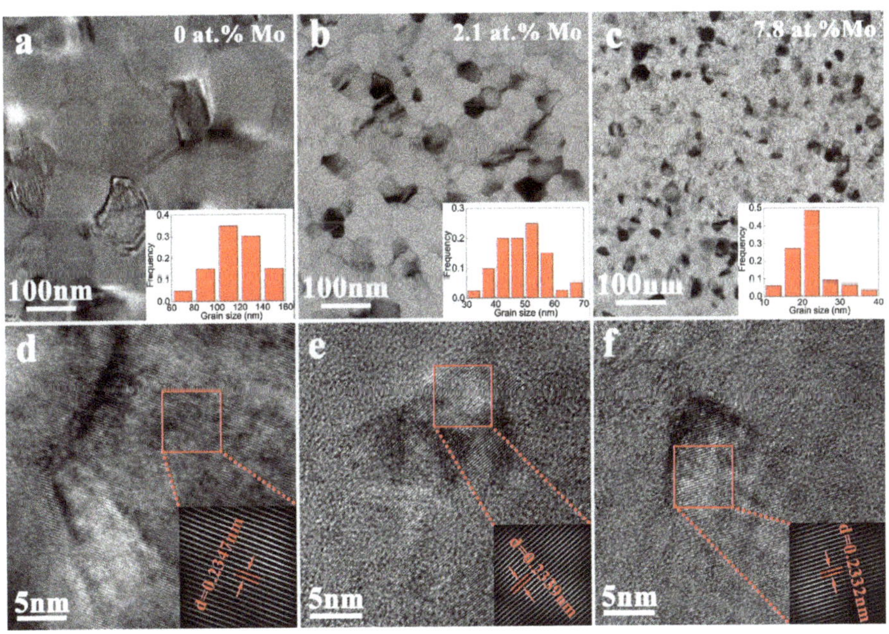

Figure 3. TEM bright field image and HRTEM image of pure Al (**a**,**d**), Al-2.1 at.% Mo (**b**,**e**), and Al-7.8 at.% Mo (**c**,**f**) alloy film.

The relationship between Mo content within the grain and the global Mo content of films is shown in Figure 4. The global Mo content is measured via EDS. The grain interior Mo content is calculated using Formula (1) [26]:

$$a = G\{2[cR_1(1) + (1-c)R_2(1)] - \beta lg[cn_1^C + (1-c)n_2^C]\} \quad (1)$$

where a is the lattice constant shown in Figure 2b. G and E are constants (for FCC structure, $G = \sqrt{2}$, $E = \frac{2-\sqrt{2}}{2}$); A and B are the numbers of the shortest and second shortest bonds, respectively. $R_1(1)$ and $R_2(1)$ are half of the bond length. n_1^C and n_2^C are the numbers of covalent electrons.

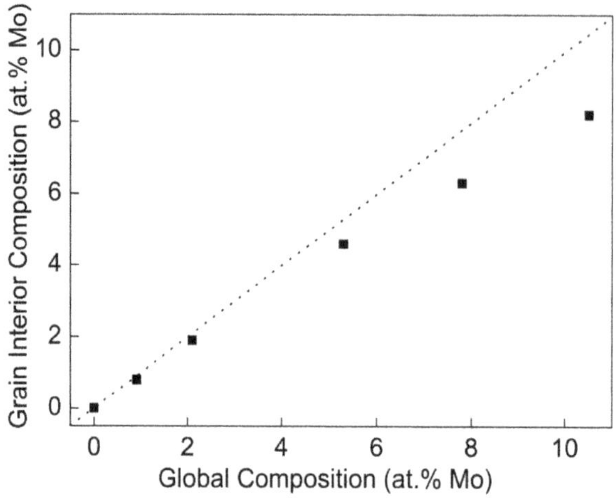

Figure 4. The variations in the grain interior Mo content and global Mo content of Al-Mo films.

In Figure 4, the dashed line shows that the grain interior's Mo content is equal to the global Mo content. As the global Mo content increases, the grain interior's Mo content also shows an upward trend, but is always generally lower than the global Mo content. This indicates the phenomenon of Mo atom segregation at grain boundaries in Al-Mo alloy films. When the total Mo content is below 2.1 at.%, this difference is not significant, but when the Mo content is above 5.3 at.%, this trend gradually increases.

3.2. Mechanical Behaviors of Alloy Films

Figure 5 shows the nanoindentation measurement results of the Al-Mo alloy film. As shown in Figure 5a, the change in hardness can be divided into two stages with the increase in Mo content. When the Mo content is in the range of 0–2.1 at.%, the hardness rapidly increases from 2.2 GPa to 3.7 GPa. When the Mo content is over 2.1 at.%, although the hardness continues to increase, the growth rate gradually slows down. When the Mo content increases to 10.5 at.%, the hardness reaches the maximum value of 4.9 GPa. The elastic modulus in Figure 5b shows an almost linear increasing trend with the increase in Mo content. The dashed line is the fitting curve.

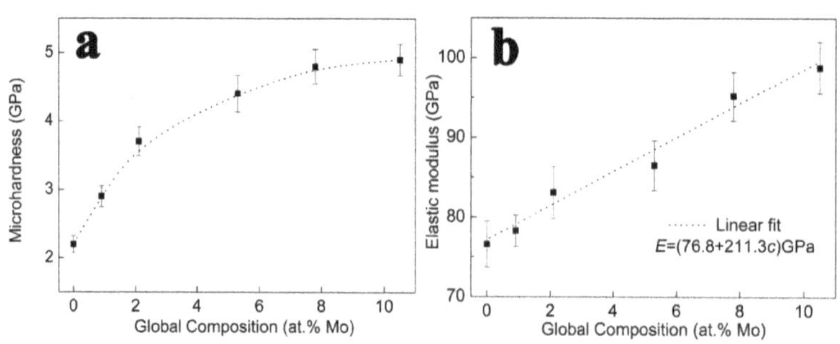

Figure 5. The nanoindentation measurement results of Al-Mo alloy film's (**a**) hardness and (**b**) elastic modulus.

There are three strengthening factors that contribute to the increase in the hardness of alloy films. On the one hand, as the Mo content increases, the grain size of the alloy film continues to decrease. The hardness increment of ΔH_{gb} caused by grain refinement is one of the important reasons for the increase in the hardness of Al-Mo alloy films.

The value of ΔH_{gb} was calculated using Formula (2). In this article, ΔH_{gb} represents the difference in hardness between Al-Mo alloy films with different Mo contents and pure Al films.

$$\Delta H_{gb} = H - H_{Al} = ak(d^{-1/2} - d_{Al}^{-1/2}) \tag{2}$$

where a = 3.6 represents the conversion relationship between hardness and strength. The value of k is taken as 3.795 $\mathrm{GPa/nm^{-1/2}}$ [18].

On the other hand, solid-solution atoms also have pinning and hindering effects on the movement of dislocations. Therefore, solid-solution strengthening is another important reason for the enhancement of the hardness of Al-Mo alloy thin films. The hardness increment ΔH_{ss} can be calculated using the Fleischer model.

$$\Delta \sigma_{Fleischer} = \sqrt{2}(0.2)^{3/2} G |\varepsilon_G' - m\varepsilon_b|^{3/2} c^{1/2} \tag{3}$$

where c is the solute content and m is a constant representing the type of dislocation, which is m = 3 for this paper. G is the shear modulus of the alloy film, which can be calculated from the elastic modulus in Figure 5b. ε_b is the lattice mismatch coefficient; $\varepsilon_b = (db/dc)/b$. ε_G' is the modulus mismatch coefficient.

$$\varepsilon_G' = \frac{\varepsilon_G}{1 + 0.5\varepsilon_G}$$

Finally, Rupert et al. [21] suggested that when the grain size decreases to the nanometer level, the grain boundaries also have a pinning effect on the movement of dislocations. The hardness increment $\Delta\sigma_{nc,ss}$ caused by the nanocrystalline solution pinning strengthening can be expressed as follows:

$$\Delta \sigma_{nc,ss} = \frac{Gb}{d} \varepsilon_{nc} \tag{4}$$

Based on the above, the total hardness increment ΔH_{Total} of the Al-Mo alloy film can be expressed as follows:

$$\Delta H_{Total} = \Delta H_{gb} + \Delta H_{ss} + \Delta H_{nc,\ ss} \tag{5}$$

Figure 6 shows the variation in the curves of ΔH_{Total}, ΔH_{gb}, ΔH_{ss}, and $\Delta H_{nc,ss}$ with the Mo content. As shown in the figure, the increment in hardness caused by the three strengthening factors gradually increases with the increase in Mo content. The contribution of grain refinement (ΔH_{gb}) is the most significant, exceeding 60%, followed by solid-solution strengthening (ΔH_{ss}), which accounts for about 30% of the effect, and the contribution of the nanocrystalline solution pinning ($\Delta H_{nc,ss}$) is the smallest, comprising only about 10% of the total effect.

The red line in Figure 7 represents the theoretically calculated hardness increment ΔH_C, while the black line represents the experimentally measured hardness increment ΔH_M. Although the ΔH_M curve is slightly higher than the ΔH_C curve, the trends of the two are generally similar. This indicates that the consistency between the theoretical calculation and experimental results is good. Although the calculated results are in good agreement with the experimental results, there is still room for discussion regarding this theoretical model. The Hall Petch formula and solid solution strengthening model are both derived from coarse-grained materials, and their applicability at the nanocrystalline scale needs to be discussed. Some studies have shown that when the grain size decreases to 20–30 nm, traditional dislocation models will fail. The nanocrystalline Al alloy will exhibit the inverse Hall–Petch effect and solid solution softening phenomenon. This may also be the reason the theoretical calculated hardness of Al-10.5 at.% Mo films continues to rise, while the measured hardness tends to flatten out.

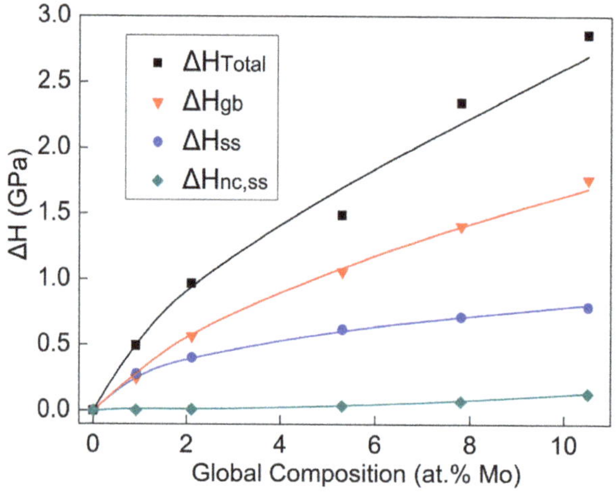

Figure 6. The variation curves of ΔH_{Total}, ΔH_{gb}, ΔH_{ss}, and $\Delta H_{nc,ss}$ with Mo content.

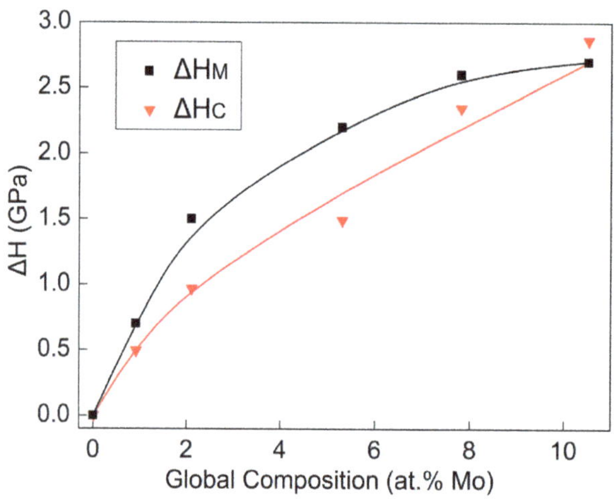

Figure 7. The calculated and measured hardness increment of alloy films.

4. Conclusions

In this work, Al-Mo alloy films were prepared using magnetron co-sputtering technology. The effect of different Mo element contents on the microstructure and comprehensive mechanical properties of the alloy films was studied. The strengthening mechanism of the alloy films was revealed. The following main conclusions were obtained:

1. All the Al-Mo alloy films formed an Al-Mo solid solution. No compounds or second phases appeared. With the increase in Mo, the grain size rapidly refines. The additional Mo atoms exhibited the phenomenon of grain boundary segregation. The higher the Mo content, the more pronounced the trend of grain boundary segregation.
2. The hardness and elastic modulus of the Al-Mo nanocrystalline alloy films gradually increased with the increase in Mo content. The fine grain (ΔH_{gb}), solid solution (ΔH_{ss}), and nanocrystalline solute pinning ($\Delta H_{nc,ss}$) are the three main reasons for the

increase in the hardness of alloy thin films. Among them, the contribution of ΔH_{gb} is the largest, accounting for over 60% of the total response, while the ΔH_{ss} accounts for about 30%, ranking second. The rest is due to the contribution of $\Delta H_{nc,ss}$.

Author Contributions: Conceptualization, Y.W. and X.Q.; methodology, Y.W. and X.Q.; validation, H.X. and Y.C.; formal analysis, Y.W. and X.Q.; investigation, Y.W. and X.Q.; writing—original draft preparation, Y.W.; writing—review and editing, X.Q. and N.Z.; visualization, Y.W. and X.Q.; supervision, N.Z.; project administration, X.Q.; funding acquisition, Y.W. and X.Q. All authors have read and agreed to the published version of the manuscript.

Funding: This work was supported by the National Natural Science Foundation of China (No. 52002240 and 51401121), Shanghai Engineering Research Center of Hot Manufacturing (18DZ2253400) and Shanghai Collaborative Innovation Center of Manufacturing Technology for Heavy Casting and Forging.

Data Availability Statement: The authors confirm that the data supporting the findings of this study are available within the article.

Conflicts of Interest: The authors declare no conflicts of interest. The funders had no role in the design of the study; in the collection, analyses, or interpretation of data; in the writing of the manuscript; or in the decision to publish the results.

References

1. Li, X.Y.; Jin, Z.H.; Zhou, X.; Lu, K. Constrained minimal-interface structures in polycrystalline copper with extremely fine grains. *Science* **2020**, *370*, 831–836. [CrossRef] [PubMed]
2. Zhou, X.; Feng, Z.; Zhu, L.; Xu, J.; Miyagi, L.; Dong, H.; Sheng, H.; Wang, Y.; Li, Q.; Ma, Y.; et al. High-pressure strengthening in ultrafine-grained metals. *Nature* **2020**, *579*, 67–72. [CrossRef]
3. Ke, X.; Ye, J.; Pan, Z.; Geng, J.; Besser, M.F.; Qu, D.; Caro, A.; Marian, J.; Ott, R.T.; Wang, Y.M.; et al. Ideal maximum strengths and defect-induced softening in nanocrystalline-nanotwinned metals. *Nat. Mater.* **2019**, *18*, 1207–1214. [CrossRef]
4. Sotniczuk, A.; Kuczyńska-Zemła, D.; Królikowski, A.; Garbacz, H. Enhancement of the corrosion resistance and mechanical properties of nanocrystalline titanium by low-temperature annealing. *Corros. Sci.* **2019**, *147*, 342–349. [CrossRef]
5. Simões, S.; Calinas, R.; Vieira, M.F.; Ferreira, P.J. In situ TEM study of grain growth in nanocrystalline copper thin films. *Nanotechnology* **2010**, *21*, 145701. [CrossRef]
6. Klement, U.; Erb, U.; El-Sherik, A.M.; Aust, K.T. Thermal stability of nanocrystalline Ni. *Mater. Sci. Eng. A* **1995**, *203*, 177–186. [CrossRef]
7. Gianola, D.; Van Petegem, S.; Legros, M.; Brandstetter, S.; Van Swygenhoven, H.; Hemker, K. Stress-assisted discontinuous grain growth and its effect on the deformation behavior of nanocrystalline aluminum thin films. *Acta Mater.* **2006**, *54*, 2253–2263. [CrossRef]
8. Malow, T.; Koch, C.C. Thermal Stability of Nanocrystalline Materials. *Mater. Sci. Forum* **1996**, *225–227*, 595–604. [CrossRef]
9. Peng, H.; Huo, W.; Zhang, W.; Tang, Y.; Zhang, S.; Huang, L.; Hou, H.; Ding, Z.; Liu, F. Correlation between stabilizing and strengthening effects due to grain boundary segregation in iron-based alloys: Theoretical models and first-principles calculations. *Acta Mater.* **2023**, *251*, 118899. [CrossRef]
10. Liang, N.; Zhao, Y. A review on thermal stability of nanostructured materials. *J. Alloys Compd.* **2023**, *938*, 168528. [CrossRef]
11. Zhou, X.; Li, X.Y.; Lu, K. Enhanced thermal stability of nanograined metals below a critical grain size. *Science* **2018**, *360*, 526–530. [CrossRef] [PubMed]
12. Devaraj, A.; Wang, W.; Vemuri, R.; Kovarik, L.; Jiang, X.; Bowden, M.; Trelewicz, J.; Mathaudhu, S.; Rohatgi, A. Grain boundary segregation and intermetallic precipitation in coarsening resistant nanocrystalline aluminum alloys. *Acta Mater.* **2019**, *165*, 698–708. [CrossRef]
13. Rajagopalan, M.; Darling, K.; Turnage, S.; Koju, R.; Hornbuckle, B.; Mishin, Y.; Solanki, K. Microstructural evolution in a nanocrystalline Cu-Ta alloy: A combined in-situ TEM and atomistic study. *Mater. Des.* **2017**, *113*, 178–185. [CrossRef]
14. Atwater, M.A.; Scattergood, R.O.; Koch, C.C. The stabilization of nanocrystalline copper by zirconium. *Mater. Sci. Eng. A* **2012**, *559*, 250–256. [CrossRef]
15. Kotan, H.; Darling, K.A.; Saber, M.; Koch, C.C.; Scattergood, R.O. Effect of zirconium on grain growth and mechanical properties of a ball-milled nanocrystalline FeNi alloy. *J. Alloys Compd.* **2012**, *551*, 621–629. [CrossRef]
16. Detor, A.J.; Schuh, C.A. Grain boundary segregation, chemical ordering and stability of nanocrystalline alloys: Atomistic computer simulations in the Ni–W system. *Acta Mater.* **2007**, *55*, 4221–4232. [CrossRef]
17. Chookajorn, T.; Murdoch, H.A.; Schuh, C.A. Design of Stable Nanocrystalline Alloys. *Science* **2012**, *337*, 951–954. [CrossRef]
18. Ma, K.; Wen, H.; Hu, T.; Topping, T.D.; Isheim, D.; Seidman, D.N.; Lavernia, E.J.; Schoenung, J.M. Mechanical behavior and strengthening mechanisms in ultrafine grain precipitation-strengthened aluminum alloy. *Acta Mater.* **2013**, *62*, 141–155. [CrossRef]

19. Scattergood, R.O.; Koch, C.C.; Murty, K.L.; Brenner, D. Strengthening mechanisms in nanocrystalline alloys. *Mater. Sci. Eng. A* **2008**, *493*, 3–11. [CrossRef]
20. Wen, H.; Topping, T.D.; Isheim, D.; Seidman, D.N.; Lavernia, E.J. Strengthening mechanisms in a high-strength bulk nanostructured Cu–Zn–Al alloy processed via cryomilling and spark plasma sintering. *Acta Mater.* **2013**, *61*, 2769–2782. [CrossRef]
21. Rupert, T.J.; Trenkle, J.C.; Schuh, C.A. Enhanced solid solution effects on the strength of nanocrystalline alloys. *Acta Mater.* **2011**, *59*, 1619–1631. [CrossRef]
22. Shanmugasundaram, T.; Heilmaier, M.; Murty, B.S.; Sarma, V.S. On the Hall–Petch relationship in a nanostructured Al–Cu alloy. *Mater. Sci. Eng. A* **2010**, *527*, 7821–7825. [CrossRef]
23. Darling, K.; Roberts, A.; Armstrong, L.; Kapoor, D.; Tschopp, M.; Kecskes, L.; Mathaudhu, S. Influence of Mn solute content on grain size reduction and improved strength in mechanically alloyed Al–Mn alloys. *Mater. Sci. Eng. A* **2013**, *589*, 57–65. [CrossRef]
24. Dong, Y.J.; Yang, D.; Shen, J.; Li, G.Y. Improved soluble salt substrate method for thin film TEM sample preparation. *Mater. Res. Innov.* **2012**, *16*, 158–160. [CrossRef]
25. Pharr, G.; Oliver, W. Measurement of Thin Film Mechanical Properties Using Nanoindentation. *MRS Bull.* **1992**, *17*, 28–33. [CrossRef]
26. Yu, R.H. The empirical electron theory of solids and molecules. *Chin. Sci. Bull.* **1978**, *23*, 217–224.

nanomaterials

MDPI

Article

Application of Polyvinyl Alcohol–Ethylene Glycol Hydrogel Technology for Removing Animal Glue in Book Restoration Based on Fluorescent Labeling Evaluation

Jia Wang [1], Yuting Xu [2], Canxin Tian [2], Yunjiang Yu [2] and Changwei Zou [2,*]

1 Library, Lingnan Normal University, Zhanjiang 524048, China; wangjia@lingnan.edu.cn
2 Department of Physics, Lingnan Normal University, Zhanjiang 524048, China; 13433129012@163.com (Y.X.); cxtian@lingnan.edu.cn (C.T.); sidui@lingnan.edu.cn (Y.Y.)
* Correspondence: zoucw@lingnan.edu.cn

Abstract: This study developed a novel material based on polyvinyl alcohol–ethylene glycol (PVA-EG) hydrogel and systematically evaluated its potential application in the removal of animal glue from book surfaces. The microstructure, surface properties, and mechanical characteristics of the PVA-EG hydrogel were analyzed using X-ray diffraction (XRD), Fourier-transform infrared spectroscopy (FTIR), contact angle measurements, a universal testing machine, and a dynamic mechanical analysis (DMA). The introduction of ethylene glycol (EG) could weaken hydrogen bonding interactions between PVA chains to enhance the molecular chain flexibility of the hydrogel. Notably, the 10% PVA-EG hydrogel shows better crystallinity, higher hydrophilicity, and optimal balance between mechanical strength and flexibility compared to pure PVA, which is conducive to improving the efficiency of the removal of animal glue. Additionally, the effectiveness of the process of removing animal glue was verified by real-time monitoring using europium nitrate at a concentration of 0.4% (w/v) as a fluorescent marker. Such hydrogels with high mechanical properties, strong surface hydrophilicity, good removal efficiency, and gentle treatment characteristics have potential applications in the restoration of cultural heritage.

Keywords: PVA-EG hydrogel; animal glue removal; book restoration; fluorescence labeling

1. Introduction

The protection of cultural heritage is of great significance to the inheritance of human civilization. Many precious ancient books, such as ancient texts, calligraphy, and paintings, and paper artifacts like murals, have developed problems of aging and deterioration over time due to the use of traditional adhesives like animal glue [1–3]. The animal glue becomes brittle over a long period, leading to damage to paper fibers and pigment layers, which poses significant challenges to the restoration of ancient books [4–6]. Traditional cleaning methods like mechanical scraping and the use of organic solvents can cause additional damage to ancient books and lack precise control over the cleaning process [7]. These methods may damage the substrate materials of ancient books or leave harmful residues, affecting their long-term preservation. There is an urgent need to develop a gentle, controllable, and non-destructive method of removing animal glue for the restoration of classical books.

In recent years, the new hydrogel cleaning system has gradually become an essential tool for the cleaning of ancient books due to its controllability and gentleness [8–10]. These hydrogels can confine cleaning agents within the gel network, achieving controllable release of the cleaning solution, and avoiding the risk of the penetration and diffusion of liquid cleaning agents, thereby reducing the damage to ancient books.

Polyvinyl alcohol (PVA)-based hydrogels have been widely used in the field of ancient book cleaning due to their biocompatibility, tunable mechanical properties, and good adhesiveness [11–14]. By blending PVA hydrogels with other natural or synthetic polymers—such as polyvinylpyrrolidone (PVP), starch, chitosan, agarose, methacrylates, etc.—or by introducing cross-linking agents, their properties can be further enhanced, including increased mechanical strength, water retention capacity, and plasticity [15–20]. Recent research has further broadened the applications of PVA hydrogels in cleaning ancient books. For example, the "green" PVA/starch cryogels prepared by introducing starch not only have excellent cleaning performance but also enhance ecological compatibility [21]. In addition, PVA-based sponge-like cryogels fabricated through freeze–thaw cycles and solvent exchange methods possess adjustable pore structures and excellent diffusion properties, which make them ideal for cleaning complex surfaces [22–24]. The optimized semi-interpenetrating p(HEMA)/PVP hydrogel accelerates the gel preparation process and enhances the effectiveness of cleaning artwork surfaces [25].

To tackle the challenge of removing animal glue, recent studies proposed the use of a heat-induced PVA hydrogel dissolution method to effectively eliminate animal glue from ancient mural surfaces [26]. This method avoids damage of the mural substrate base by partially dissolving the PVA hydrogel under heat to control the release of hot water, gradually dissolving and removing the animal glue layer. Moreover, microemulsion-loaded PVA/PEI hydrogels have also been used to clean animal glue and dirt from the surfaces of murals and ancient coins, demonstrating their potential in cleaning ancient books [27]. In addition to PVA-based hydrogels, other types of gel materials have also been used in cultural heritage cleaning. For example, the reusable cross-linked gels based on poly(ethyl methacrylate) (PEMA) with adjustable flexibility and affinity for organic solvents can be used to clean aged coatings on artwork surfaces [28]. The combination of bioenzymes and biosurfactants has been used to remove mold spots from paper artifacts, showing good cleaning effectiveness and material compatibility [29]. Polyelectrolyte hydrogels have been used for non-destructive cleaning of murals, enhancing the mechanical properties of the gels and their adsorption capacity for dyes and metal ions by introducing chitosan and charged copolymers [30]. More research is now focusing on green chemistry and renewable materials to enhance the sustainability and eco-friendliness of cleaning methods [31,32]. For instance, nanostructured bio-based organogels derived from castor oil and prepared via a sustainable polyurethane cross-linking method can be used for cleaning water-sensitive artworks [33]. Starch nanoparticles prepared through non-solvent methods using green solvents can be used to reinforce fragile painting layers [34].

Current hydrogels still present some specific limitations in practical applications, including restricted adaptability to complex surfaces, complex preparation procedures, and the possibility of gel residue [35–40]. To solve these problems, this study focuses on developing an optimized PVA-EG hydrogel by adjusting PVA concentrations (6%, 8%, 10%, 12%) and incorporating ethylene glycol (EG) as an additive to achieve hydrogels with enhanced mechanical properties and water absorption capacity. We prepared the hydrogel by the low-temperature freeze–thaw cycling and solvent replacement methods. The structure and properties of the hydrogels were characterized by XRD, FTIR, contact angle measurements, mechanical testing, and DMA. The results indicated that introducing EG facilitated the orderly arrangement of PVA molecular chains, enhancing the crystallinity and mechanical strength of the hydrogels. Fluorescently labeled animal glue was used to simulate ancient books, and the cleaning effectiveness of PVA-EG hydrogels was evaluated on glue layers of different thicknesses and we confirmed the feasibility of this approach. This optimizing method offers a new gentle, efficient, and controllable technique for cultural heritage preservation.

2. Materials and Methods

2.1. Materials

Polyvinyl alcohol (PVA, 96.0–98.0% hydrolysis degree) and ethylene glycol (EG, analytical grade) were used directly in the experiments without further purification. Fluorescent dye (europium nitrate solution) was used to simulate the book surface and to prepare fluorescent animal glue.

2.2. Preparation of PVA-EG Hydrogel

PVA-EG hydrogels were prepared by the low-temperature freeze–thaw cycling method and solvent exchange technique. First, PVA particles of different concentrations (6%, 8%, 10%, 12%) were dissolved in a mixture of deionized water and ethylene glycol (mass ratio of 4:6), and stirred continuously at 90 °C until the PVA was fully dissolved. The solution was then poured into Petri dishes and frozen at −24 °C for 24 h, followed by thawing at room temperature for 4 h. To guarantee the formation of a stable hydrogel structure, the freeze–thaw process was repeated three times. A 10% PVA hydrogel was prepared following the same method. After the freeze–thaw cycles, the PVA-EG hydrogel was immersed in deionized water for 24 h, changing the water every 3 h to replace the EG.

2.3. Characterization Techniques

The crystalline structure of PVA hydrogel samples was analyzed using an X-ray diffractometer (Empyrean, PANalytical, Almelo, The Netherlands). The testing range was from 10° to 80° (2θ) with a scanning speed of 2°/min, using a Cu K_α radiation source (λ = 1.5406 Å). The FTIR spectra of the hydrogel samples were recorded using a Fourier-transform infrared spectrometer (IRPrestige-21, Shimadzu, Kyoto, Japan) over a scanning range of 3750–750 cm^{-1}, with an attenuated total reflectance (ATR) accessory. The ATR accessory used a zinc selenide (ZnSe) crystal to ensure good contact with the sample surface and obtain high-quality spectral data. The surface wettability and hydrophilicity of PVA hydrogel samples with different concentrations were tested using a contact angle measuring instrument (DSA100, Krüss GmbH, Hamburg, Germany). Each hydrogel sample was tested three times to ensure the repeatability of the measurements, thereby evaluating the surface wettability and hydrophilicity of the samples. Tensile tests of the PVA hydrogel samples were conducted using a universal testing machine (Instron 3365, Instron Corporation, Boston, MA, USA) in accordance with ASTM D638 standards. The sample dimensions were 12 mm in width, 50 mm in length, and 5 mm in thickness, and the tensile test loading speed was set to 5 mm/min. Each sample was tested three times, and the average values were recorded. The viscoelastic properties of the PVA hydrogel samples were tested in the shear mode using a dynamic mechanical analyzer (DMA 242 E Artemis, Netzsch, Stuttgart, Germany). The test conditions included a frequency of 1 Hz and a temperature range from 30 °C to 50 °C, with a heating rate of 2 °C/min.

2.4. Fluorescent Animal Glue Removal Experiment

2.4.1. Preparation of Fluorescent Animal Glue

To simulate the animal glue layers found in book restoration and to determine the optimal concentration of a fluorescent marker, experiments were conducted to evaluate the effect of various concentrations of europium nitrate ($Eu(NO_3)_3$) on fluorescence intensity. In total, 10 g of animal glue granules was added to 90 mL of deionized water and heated to 60 °C in a water bath with continuous stirring until fully dissolved. Then, the europium nitrate solution was added to the animal glue solution to prepare 0.1%, 0.2%, 0.4%, 0.6%, 0.8%, and 1.0% (*w*/*v*) europium nitrate solutions. The fluorescence spectra of the europium nitrate animal glue solutions at different concentrations were measured using a fluorescence spectrophotometer. In this study, fluorescence measurements were conducted using a steady-state and transient fluorescence spectrometer (FLS1000, Edinburgh Instruments, Livingston, UK), with an excitation wavelength set at 318 nm and an emission scanning range of 550–750 nm.

2.4.2. Preparation of Simulated Samples

The prepared europium nitrate fluorescent animal glue solution was evenly coated onto 5 cm × 5 cm filter paper with thicknesses of 5 μm, 10 μm, 20 μm, and 50 μm by using a blade coater. After coating, the samples were dried at 25 °C in a constant-temperature and -humidity chamber for 24 h to ensure full curing of the glue layer.

2.4.3. Animal Glue Removal Experiment on Simulated Samples

PVA hydrogels were cut into 5 cm × 5 cm sheets with a thickness of 5 mm and placed over the fluorescent animal glue layer. An infrared heating lamp (150 W) was placed about 10 cm above the hydrogel, and the temperature was controlled at 45 °C. The removal time was adjusted according to the thickness of the glue layer. At each removal time point, the fluorescence spectrum of the sample was measured with a fluorescence spectrophotometer, and the removal efficiency was assessed based on changes in fluorescence intensity.

2.4.4. Animal Glue Removal Experiment on Book Samples

To verify the feasibility of PVA hydrogels in real-world book restoration, we selected a page adhered with animal glue for the experiment. The thickness of the animal glue layer on the book page was measured using a high-precision thickness gauge, and the appropriate removal method was selected based on the thickness. The method followed the fluorescent animal glue removal experiment. At each time point, the surface of the book page was measured using a Fourier-transform infrared spectrometer (FTIR) to evaluate the removal effectiveness.

3. Results and Discussion

3.1. Crystal Structure Analysis

Figure 1 shows the changes in diffraction peaks of PVA-EG hydrogel samples with different concentrations. All samples exhibited a distinct diffraction peak at around 19.4°, corresponding to the (101) crystal plane of PVA, indicating that the introduction of EG did not disrupt the crystal structure of PVA [41]. In contrast, the addition of EG effectively promoted the orderly arrangement of PVA segments, further improving the crystalline structure of the samples. Notably, in the 10% and 12% PVA-EG samples, the intensity and sharpness of the diffraction peaks were significantly enhanced compared to other samples. This can be attributed to the formation of additional hydrogen bonds due to the presence of EG, which strengthened the intermolecular interactions between PVA chains, thereby promoting a more ordered molecular arrangement.

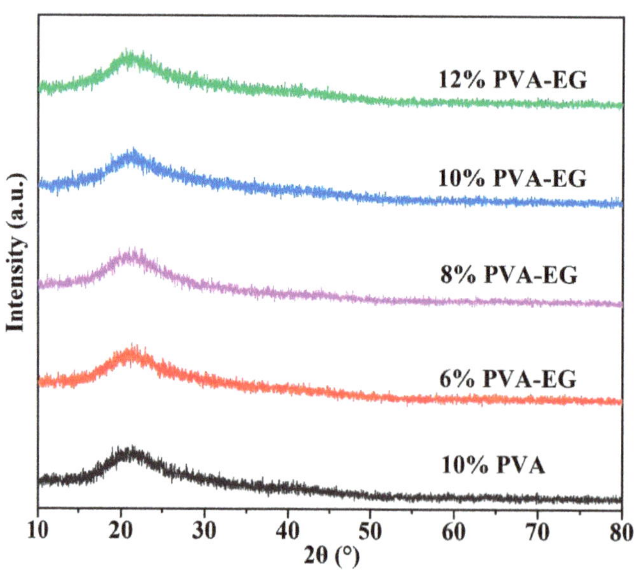

Figure 1. XRD patterns of PVA-EG hydrogel samples at various concentrations.

3.2. Chemical Structure Characterization

Figure 2 illustrates the changes in FTIR absorption peaks of PVA-EG hydrogel samples at various concentrations. The results show that the O-H stretching vibration peak (3300 cm^{-1}) was gradually broadened and the intensity was decreased as the concentration of EG increased. This suggests that the addition of EG can weaken the hydrogen bonding interactions between PVA molecular chains by inserting itself between the PVA segments. The O-H peak in the pure PVA sample is more distinct, indicating a relatively dense hydrogen bond network [42]. In samples with higher PVA content, more PVA molecular chains can form hydrogen bonds with EG molecules, which partially inhibits the complete hydrolysis of PVA and retains the acetyl (-O-C=O) groups; however, the change in the C=O signal was not significant [43]. Additionally, changes in the absorption peak were observed around 1700 cm^{-1}. According to the study of Tretinnikov, the carbonyl absorption peak in PVA can reflect changes in molecular arrangement. One of the peaks in the doublet may correspond to the carbonyl groups in PVA molecules, while the other may result from interactions between EG and PVA, possibly forming a new hydrogen bond network and leading to a more ordered arrangement of PVA segments [42].

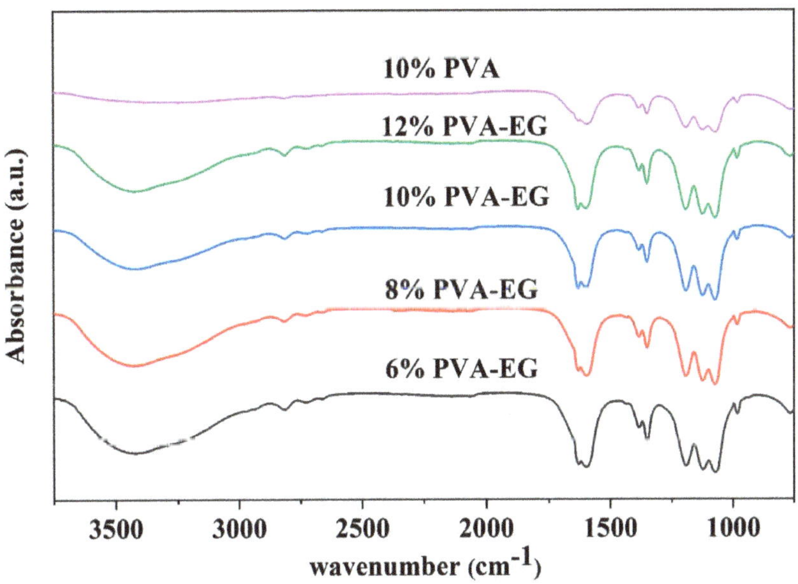

Figure 2. FTIR analysis of PVA-EG hydrogels at different concentrations.

3.3. Surface Wettability Analysis

Figure 3 illustrates the changes in the wettability of PVA-EG hydrogels at various concentrations. The pure PVA sample had a larger initial contact angle, and the contact angle decreased slowly over time, indicating poor surface wettability and relatively weak water absorption capacity. This could be attributed to the stronger hydrogen bonding between pure PVA molecular chains, causing tighter molecular packing and less exposure of hydrophilic groups, thus reducing interaction with water molecules. With the increase in the EG, the contact angle of the hydrogel gradually decreased, and the wettability significantly improved. Particularly in the 10% and 12% PVA-EG samples, the water droplets spread significantly at 0.1 s and were almost fully absorbed after 0.4 s with the contact angle approaching 0. This indicates that the introduction of EG significantly improved the hydrophilicity and water absorption capacity of the PVA hydrogel. EG molecules formed new hydrogen bonds with PVA molecular chains, increasing the polarity of the hydrogel and the density of hydrophilic groups. This change in structure increased the surface free energy of the hydrogel, facilitating interactions with water molecules and accelerating water absorption and penetration. During heating, the hydrogel with excellent water absorption can release more moisture, maintaining the glue layer in a hydrated state and further improving the removal efficiency.

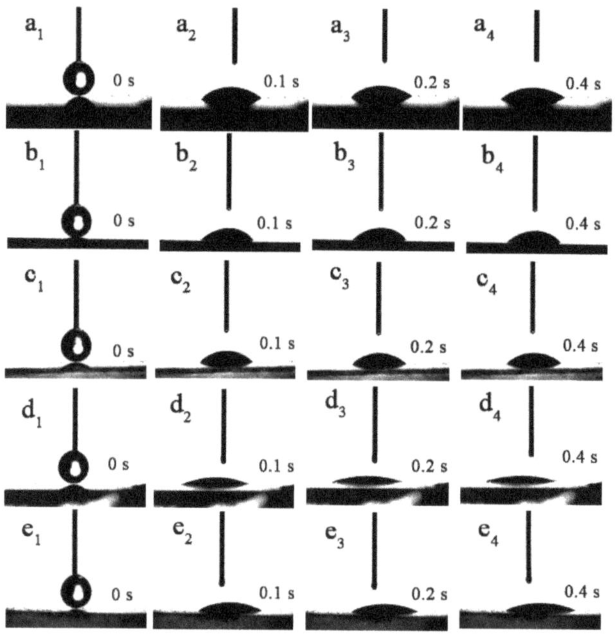

Figure 3. Contact angle measurement results of PVA-EG hydrogel samples at different concentrations: (**a**) 10% pure PVA hydrogel, (**b**) 6% PVA-EG hydrogel, (**c**) 8% PVA-EG hydrogel, (**d**) 10% PVA-EG hydrogel, (**e**) 12% PVA-EG hydrogel.

3.4. Mechanical Property Testing

Figure 4 illustrates the stress–strain curves of PVA-EG hydrogels with various concentrations. The results indicate that the mechanical properties of the hydrogel significantly improve with the addition of EG, particularly in samples with higher PVA concentrations. Specifically, the 6% PVA-EG hydrogel has good flexibility with a maximum strain of about 150%, but its mechanical strength is low with a maximum stress of only 0.05 $\pm$ 0.003 MPa, limiting its use in applications requiring mechanical support. The 8% PVA-EG hydrogel strikes a good balance between rigidity and flexibility, with a maximum strain of 120% and a maximum stress of 0.08 $\pm$ 0.005 MPa, showing significant improvement over the 6% concentration. For the 10% PVA-EG hydrogel, its mechanical strength was notably improved, reaching a maximum stress of 0.12 $\pm$ 0.009 MPa while maintaining a maximum strain of around 110%. This balance between strength and ductility allows it to effectively tackle stubborn animal glue without causing excessive stress concentration on the book surface. The 12% PVA-EG hydrogel exhibited the strongest mechanical strength, with a maximum stress approaching 0.15 $\pm$ 0.011 MPa and a strain of 150%, demonstrating excellent strength and ductility. However, its relatively low flexibility could pose a risk of damaging fragile surfaces when working with delicate substrates. The 10% PVA-EG hydrogel demonstrated the best mechanical properties, providing sufficient strength to remove stubborn animal glue while maintaining moderate flexibility, effectively meeting various complex book restoration needs.

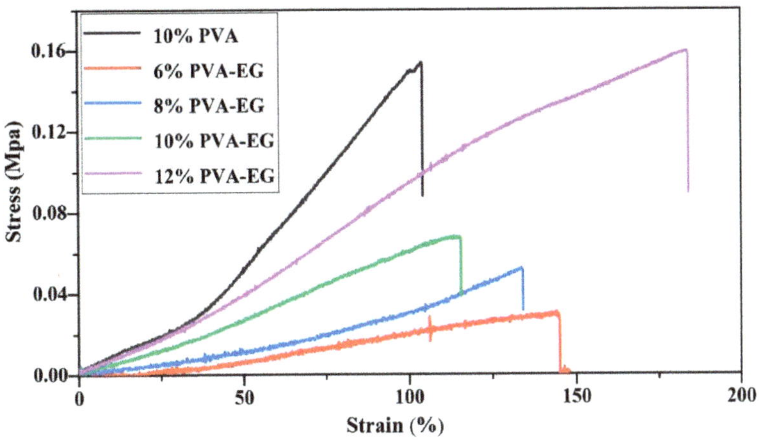

Figure 4. Stress–strain curves of PVA-EG hydrogels at different concentrations.

3.5. Dynamic Mechanical Analysis (DMA)

Figure 5 shows the viscoelastic behavior of PVA-EG hydrogels at different concentrations in the temperature range of 30 °C to 50 °C. The 6% PVA-EG hydrogel exhibited the lowest E′ value across the entire temperature range, especially when the temperature exceeded 45 °C, where E′ dropped to 0.003 ± 0.0001 MPa, indicating lower elasticity and insufficient structural rigidity. In contrast, the E′ value of the 8% PVA-EG hydrogel was 0.009 ± 0.0004 MPa at 45 °C, indicating the elasticity with slight improvement compared to the 6% PVA-EG. The tan δ value was 0.05, suggesting a better balance with viscoelasticity, but the elasticity at high temperatures was still low, making it suitable for more moderate removal applications. The 10% PVA-EG hydrogel exhibited a relatively high and stable E′ value from 30 °C to 50 °C with an E′ value of 0.012 ± 0.0005 MPa at 45 °C, indicating excellent elasticity and mechanical performance. The tan δ value remained below 0.10, indicating a good balance between viscosity and elasticity with minimal energy dissipation. The E′ value of the 12% PVA-EG hydrogel was 0.013 ± 0.0005 MPa at 45 °C, showing relatively good elasticity, but its tan δ value reached 0.30, indicating a significant increase in the viscous component and energy dissipation, which may lead to a decrease in mechanical performance during high-temperature operations. The 10% PVA-EG hydrogel achieved an optimal balance between the storage modulus and loss factor, combining high elasticity with low energy dissipation, making it suitable for applications requiring structural stability and mechanical performance under high-temperature conditions, especially in the removal of animal glue.

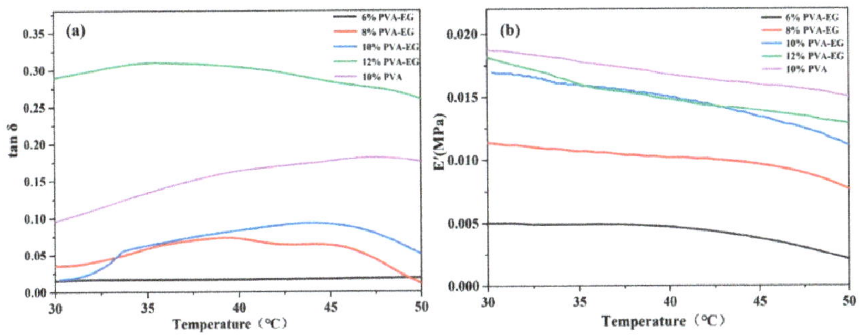

Figure 5. Viscoelastic behavior of PVA-EG hydrogels at different concentrations in the temperature range of 30 °C to 50 °C. (**a**) Loss factor tan δ of PVA-EG hydrogels; (**b**) Storage modulus E′ of PVA-EG hydrogels.

3.6. Analysis of Fluorescent Animal Glue Removaling Results

3.6.1. Optimization of Europium Nitrate Concentration and Its Effect on Fluorescent Labeling

Figure 6 illustrates the impact of various europium nitrate concentrations on the fluorescence intensity of animal glue. The results show that the fluorescence intensity first increased and then decreased with the increase in europium nitrate concentration. The fluorescence intensity peaked at a concentration of 0.4%. When the concentration exceeded 0.4%, the fluorescence intensity dropped significantly, indicating a concentration quenching effect. The position of the fluorescence emission peak remained stable at all concentrations, indicating that the concentration changes primarily affected fluorescence intensity, not emission wavelength. Therefore, 0.4% (*w*/*v*) europium nitrate was selected as the optimal concentration for fluorescent labeling, providing a strong and stable fluorescence signal for subsequent experiments.

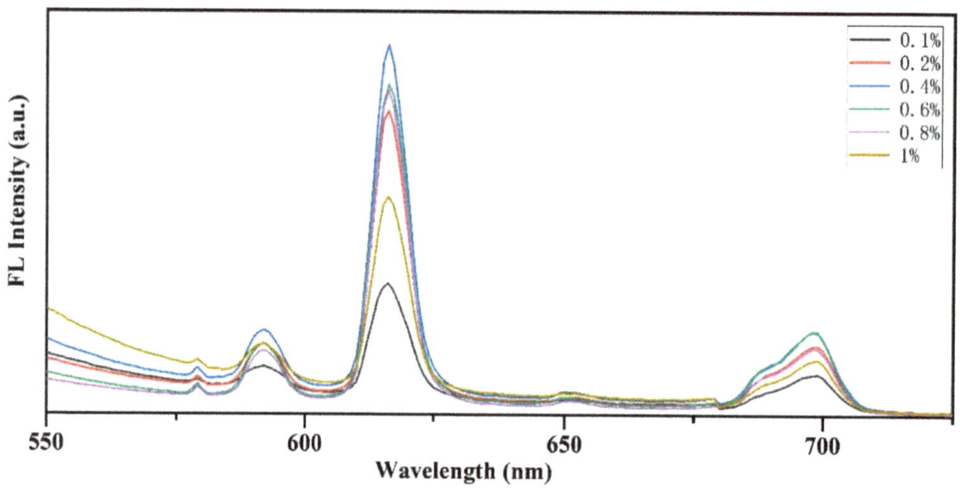

Figure 6. Effect of different europium nitrate concentrations on fluorescence intensity of animal glue.

3.6.2. Removal Efficiency of Simulated Animal Glue Layers by PVA-EG Hydrogel

Figure 7 shows the removal efficiency of PVA-EG hydrogel on fluorescent-labeled animal glue of different thicknesses. The results show that the fluorescence intensity rapidly decreased for thin glue layers (5 and 10 μm), reaching baseline levels within 12 min and 20 min, respectively, indicating that the glue layers were effectively removed. This indicates that the water in the hydrogel can quickly penetrate, soften, and dissolve the thin glue layers. For medium-thickness glue layers (20 μm), the fluorescence intensity significantly decreased within 24 min, and the fluorescence signal was almost eliminated, suggesting that most of the animal glue had been removed. For thicker glue layers (50 μm), the fluorescence intensity dropped significantly, nearing baseline levels after 36 min of treatment, indicating that most of the animal glue had been removed. These experiments demonstrate that the PVA-EG hydrogel is effective in removing animal glue layers of various thicknesses with the removal efficiency inversely proportional to the thickness of the glue layer.

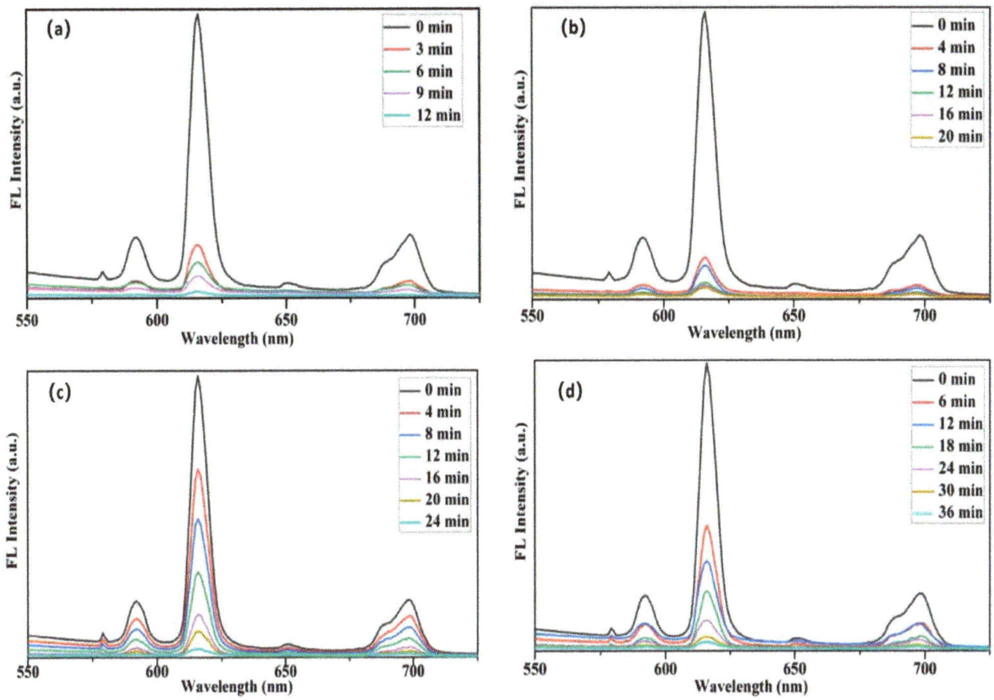

Figure 7. Removal efficiency of PVA-EG hydrogel on fluorescent-labeled animal glue layers of different thicknesses. (**a**) Fluorescence intensity changes for the 5 μm glue layer; (**b**) Fluorescence intensity changes for the 10 μm glue layer; (**c**) Fluorescence intensity changes for the 20 μm glue layer; (**d**) Fluorescence intensity changes for the 50 μm glue layer.

3.6.3. Application of PVA-EG Hydrogel in Removal of Animal Glue from Book Pages

To assess the practical effectiveness of the PVA-EG hydrogel, we applied it to remove a 35 μm thick animal glue layer from the pages of a book. During the removal process, we used FTIR to monitor the change in the glue layer, focusing on the characteristic absorption peaks of the animal glue—amide I (1650 cm^{-1}) and amide II (1550 cm^{-1}). Figure 8 illustrates the effectiveness of the PVA-EG hydrogel in removing animal glue from book pages. The results showed that the intensity of the amide peak gradually weakened and disappeared with the increase in treatment time. After 12 min of treatment, the intensity of the amide peak decreased significantly, indicating that part of the adhesive layer had been removed. At 24 min, the amide peak disappeared almost completely, indicating that most of the animal glue had been removed. After 30 min, the amide peak had completely disappeared, and the infrared spectrum matched that of the untreated paper substrate, confirming that the adhesive layer had been fully removed. These results demonstrate that PVA-EG hydrogels can efficiently remove animal glue from book pages within a relatively short time, and the removal efficiency improves with increasing treatment time.

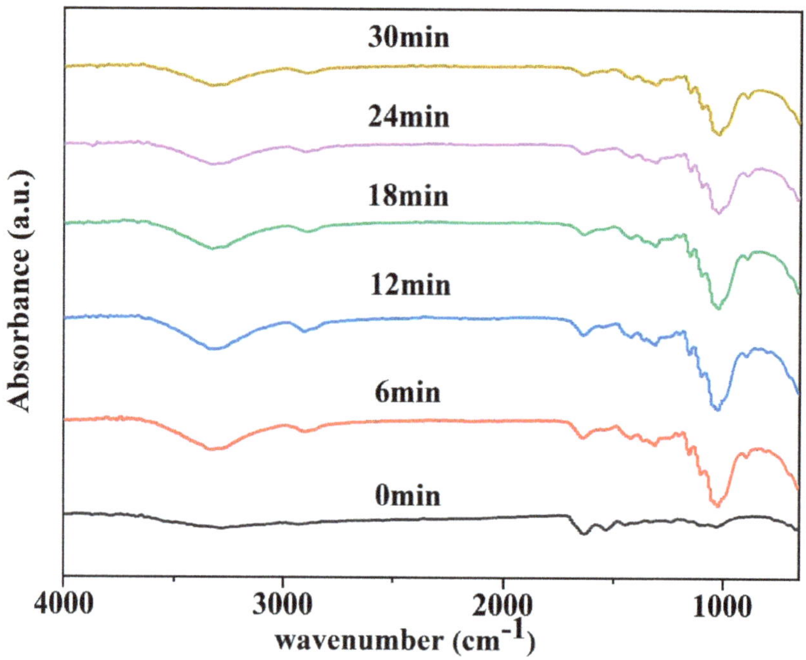

Figure 8. The effect of PVA-EG hydrogel on the removal of animal glue from book pages.

3.6.4. Mechanism of Animal Glue Removal by PVA-EG Hydrogel

From a theoretical perspective, the mechanism of efficient animal glue removal by PVA-EG hydrogel can be attributed to the synergistic effect of its molecular structure and interfacial physicochemical properties. The addition of EG plays a key role in the hydrogel system by regulating hydrogen bonding between PVA molecular chains, which significantly improves the crystallinity and mechanical properties of the material. EG enhances the flexibility of PVA molecular chains by forming new hydrogen bonds and enables the hydrogel to conform to complex surfaces, ensuring close coverage of the animal glue surface during application. The addition of EG also significantly improved the hydrophilicity of the hydrogel, enhancing its water transfer capacity. This characteristic allows water to quickly penetrate the animal glue layer, promoting its hydrolysis and softening. The mechanical stability of the hydrogel structure was strengthened by the regulation of EG, enabling the PVA-EG hydrogel to maintain prolonged effective contact during operation, ensuring the continuity and thoroughness of the removal process.

4. Conclusions

This study employed various methods, including XRD, FTIR, contact angle measurements, a universal tensile testing machine, and DMA, to explore the potential application of PVA-EG hydrogels in gently removing animal glue from book surfaces. The addition of EG improved both the crystallinity and mechanical properties of the hydrogel by modulating hydrogen bonding interactions among PVA molecular chains. The XRD analysis revealed that as crystallinity increased, the crystal structure of PVA-EG hydrogels became more organized. Notably, the 10% PVA-EG hydrogel achieved an optimal balance between mechanical strength and flexibility. The FTIR analysis further illustrated that EG facilitated rearrangement and flexibility within the molecular chains, thereby enhancing both mechanical stability and removal efficiency of the hydrogel. A combined analysis using DMA tests and stress–strain curves demonstrated that PVA-EG hydrogels possess excellent energy storage capacity and viscoelastic balance across varying temperatures.

Specifically, the 10% PVA-EG hydrogel maintained stable elasticity and strength even at elevated temperatures while also exhibiting commendable extensibility with low energy dissipation. Overall, our results suggest that the PVA-EG hydrogel can effectively yet gently remove glue layers in a relatively short time frame. The introduction of EG into this formulation successfully achieved a harmonious balance among removal efficiency, mechanical performance, and viscoelasticity.

Author Contributions: Formal analysis, J.W. and C.Z.; Investigation, Y.X., C.T., Y.Y. and C.Z.; Resources, Y.Y.; Data curation, J.W., C.T. and Y.Y; Writing—original draft, J.W. and Y.Y.; Funding acquisition, C.Z. All authors have read and agreed to the published version of the manuscript.

Funding: This work was supported by the Innovation team project of Guangdong Universities (2020KCXTD032) and Lingnan Normal University (LT2203), and the Science and Technology Project of Zhanjiang (2022A0100, 2022B01044, 2024B01117).

Data Availability Statement: Data are contained within the article.

Conflicts of Interest: The authors declare no conflict of interest.

References

1. Domingues, J.A.; Bonelli, N.; Giorgi, R.; Fratini, E.; Gorel, F.; Baglioni, P. Innovative hydrogels based on semi-interpenetrating p (HEMA)/PVP networks for the cleaning of water-sensitive cultural heritage artifacts. *Langmuir* **2013**, *29*, 2746–2755. [CrossRef] [PubMed]
2. Bonelli, N.; Poggi, G.; Chelazzi, D.; Giorgi, R.; Baglioni, P. Poly (vinyl alcohol)/poly (vinyl pyrrolidone) hydrogels for the cleaning of art. *J. Colloid Interface Sci.* **2019**, *536*, 339–348. [CrossRef] [PubMed]
3. Mastrangelo, R.; Chelazzi, D.; Poggi, G.; Fratini, E.; Buemi, L.P.; Petruzzellis, M.L.; Baglioni, P. Twin-chain polymer hydrogels based on poly (vinyl alcohol) as new advanced tool for the cleaning of modern and contemporary art. *Proc. Natl. Acad. Sci. USA* **2020**, *117*, 7011–7020. [CrossRef] [PubMed]
4. Sun, Y.; Xiang, N.; Jiang, X.; Hou, L. Preparation of high tough poly (vinyl alcohol) hydrogel by soaking in NaCl aqueous solution. *Mater. Lett.* **2017**, *194*, 34–37. [CrossRef]
5. Baglioni, P.; Chelazzi, D. How science can contribute to the remedial conservation of cultural heritage. *Chem. A Eur. J.* **2021**, *27*, 10798–10806. [CrossRef]
6. Rosciardi, V.; Chelazzi, D.; Baglioni, P. "Green" biocomposite poly (vinyl alcohol)/starch cryogels as new advanced tools for the cleaning of artifacts. *J. Colloid Interface Sci.* **2022**, *613*, 697–708. [CrossRef]
7. Chelazzi, D.; Bordes, R.; Giorgi, R.; Holmberg, K.; Baglioni, P. The use of surfactants in the cleaning of works of art. *Curr. Opin. Colloid Interface Sci.* **2020**, *45*, 108–123. [CrossRef]
8. Cardaba, I.; Poggi, G.; Baglioni, M.; Chelazzi, D.; Maguregui, I.; Giorgi, R. Assessment of aqueous cleaning of acrylic paints using innovative cryogels. *Microchem. J.* **2020**, *152*, 104311. [CrossRef]
9. Altobelli, A.; Cennamo, P.; Trojsi, G.; Lumaga, M.R.B.; Carpentieri, A.; Fatigati, G. Experimentation of a PVA-Borax hydrogel for the removal of Paraloid B72® from artifacts of archaeological interest from the National Archaeological Museum in Naples, Italy. *Acta IMEKO* **2023**, *12*, 1–8. [CrossRef]
10. Rafiei-Sarmazdeh, Z.; Sheikh, N. Irradiation-assisted synthesis of smart hydrogels based on nanomagnetic semi-interpenetrating p (HEMA)/PVP networks for the cleaning of cultural heritage artifacts. *Discov. Appl. Sci.* **2024**, *6*, 285. [CrossRef]
11. Mazzuca, C.; Severini, L.; Domenici, F.; Toumia, Y.; Mazzotta, F.; Micheli, L.; Titubante, M.; Di Napoli, B.; Paradossi, G.; Palleschi, A. Polyvinyl alcohol based hydrogels as new tunable materials for application in the cultural heritage field. *Colloids Surf. B Biointerfaces* **2020**, *188*, 110777. [CrossRef] [PubMed]
12. Stagno, V.; Genova, C.; Zoratto, N.; Favero, G.; Capuani, S. Single-sided portable NMR investigation to assess and monitor cleaning action of PVA-borax hydrogel in travertine and Lecce stone. *Molecules* **2021**, *26*, 3697. [CrossRef] [PubMed]
13. Duncan, T.T.; Chan, E.P.; Beers, K.L. Quantifying the 'press and peel'removal of particulates using elastomers and gels. *J. Cult. Herit.* **2021**, *48*, 236–243. [CrossRef]
14. Soto-Bustamante, F.; Bassu, G.; Fratini, E.; Laurati, M. Effect of composition and freeze-thaw on the network structure, porosity and mechanical properties of polyvinyl-alcohol/chitosan hydrogels. *Gels* **2023**, *9*, 396. [CrossRef] [PubMed]
15. Al-Emam, E.; Soenen, H.; Caen, J.; Janssens, K. Characterization of polyvinyl alcohol-borax/agarose (PVA-B/AG) double network hydrogel utilized for the cleaning of works of art. *Herit. Sci.* **2020**, *8*, 106. [CrossRef]
16. Wang, Y.; Li, J.; Muhammad, N.; Wang, Z.; Wu, D. Hierarchical networks of anisotropic hydrogels based on cross-linked Poly (vinyl alcohol)/Poly (vinylpyrrolidone). *Polymer* **2022**, *251*, 124920. [CrossRef]
17. Zhang, W.; Wang, Y.; Wu, D. Mapping hierarchical networks of poly (vinyl alcohol)/cellulose nanofiber composite hydrogels via viscoelastic probes. *Carbohydr. Polym.* **2022**, *288*, 119372. [CrossRef]

18. Rosciardi, V. "Green" Poly (Vinyl Alcohol)/Starch Based Cryogels for the Cleaning of Works of Art: Application, Characterization, and Investigation of the Amylose/Amylopectin Structural Role. 2022. Available online: https://flore.unifi.it/handle/2158/1291291 (accessed on 24 November 2022).
19. Severini, L.; Tavagnacco, L.; Angelini, R.; Franco, S.; Bertoldo, M.; Calosi, M.; Micheli, L.; Sennato, S.; Chiessi, E.; Ruzicka, B.; et al. Methacrylated gellan gum hydrogel: A smart tool to face complex problems in the cleaning of paper materials. *Cellulose* **2023**, *30*, 10469–10485. [CrossRef]
20. Jiang, C.; Chao, Y.; Xie, W.; Wu, D. Using bacterial cellulose to bridge covalent and physical crosslinks in hydrogels for fabricating multimodal sensors. *Int. J. Biol. Macromol.* **2024**, *263*, 130178. [CrossRef]
21. Warfield, M. Optimal PVA/Starch Complex for Cryogel Development Intended for Artifact Cleaning. 2024. Available online: https://digitalcommons.liberty.edu/research_symp/2024/posters/86/ (accessed on 16 April 2024).
22. Wu, M.; Chen, X.; Xu, J.; Zhang, H. Freeze-thaw and solvent-exchange strategy to generate physically cross-linked organogels and hydrogels of curdlan with tunable mechanical properties. *Carbohydr. Polym.* **2022**, *278*, 119003. [CrossRef]
23. Mastrangelo, R.; Resta, C.; Carretti, E.; Fratini, E.; Baglioni, P. Sponge-like Cryogels from Liquid-Liquid Phase Separation: Structure, Porosity, and Diffusional Gel Properties. *ACS Appl. Mater. Interfaces* **2023**, *15*, 46428–46439. [CrossRef] [PubMed]
24. Wu, Y.; Xing, W.; Wen, J.; Wu, Z.; Zhang, Y.; Zhang, H.; Wu, H.; Yao, H.; Xue, H.; Gao, J. Mixed solvent exchange enabled high-performance polymeric gels. *Polymer* **2023**, *267*, 125661. [CrossRef]
25. Tamburini, G.; Canevali, C.; Ferrario, S.; Bianchi, A.; Sansonetti, A.; Simonutti, R. Optimized semi-interpenetrated p (HEMA)/PVP hydrogels for artistic surface cleaning. *Materials* **2022**, *15*, 6739. [CrossRef] [PubMed]
26. Zhu, J.; Wang, J.; Wang, J.; Ding, J.; Zhang, P.; Dong, W.; Zhao, X.; Lu, Z.; Li, X. Effectively removing animal glue coated on the surface of ancient mural via dissolution of PVA hydrogel induced by thermal treatment. *J. Cult. Herit.* **2022**, *55*, 179–184. [CrossRef]
27. Zhu, J.; Wang, J.; Wang, J.; Ding, J.; Zhao, X.; Dong, W.; Lu, Z.; Li, X. Preparing a microemulsion-loaded hydrogel for cleaning wall paintings and coins. *Herit. Sci.* **2024**, *12*, 149. [CrossRef]
28. Zuliani, A.; Chen, S.; Giorgi, R. Re-usable cross-linked poly (ethyl methacrylate) gels for cleaning purposes of artworks. *Appl. Mater. Today* **2023**, *30*, 101716. [CrossRef]
29. Meng, Q.; Li, X.; Geng, J.; Liu, C.; Ben, S. A biological cleaning agent for removing mold stains from paper artifacts. *Herit. Sci.* **2023**, *11*, 243. [CrossRef]
30. Li, Q.; Wu, C.; Peng, Y.; Zhang, B. Toward a non-invasive cleaning of the wall painting using polyelectrolyte hydrogel. *Sci. China Technol. Sci.* **2023**, *66*, 2213–2224. [CrossRef]
31. Casini, A.; Chelazzi, D.; Baglioni, P. Advanced methodologies for the cleaning of works of art. *Sci. China Technol. Sci.* **2023**, *66*, 2162–2182. [CrossRef]
32. Chelazzi, D.; Baglioni, P. From nanoparticles to gels: A breakthrough in art conservation science. *Langmuir* **2023**, *39*, 10744–10755. [CrossRef]
33. Poggi, G.; Santan, H.D.; Smets, J.; Chelazzi, D.; Noferini, D.; Petruzzellis, M.L.; Buemi, L.P.; Fratini, E.; Baglioni, P. Nanostructured bio-based castor oil organogels for the cleaning of artworks. *J. Colloid Interface Sci.* **2023**, *638*, 363–374. [CrossRef] [PubMed]
34. Casini, A.; Casagli, M.; Poggi, G.; Chelazzi, D.; Baglioni, P. Tuning Local Order in Starch Nanoparticles Exploiting Nonsolvency with "Green" Solvents. *ACS Appl. Mater. Interfaces* **2024**, *16*, 21185–21196. [CrossRef] [PubMed]
35. Pensabene Buemi, L.; Petruzzellis, M.L.; Chelazzi, D.; Baglioni, M.; Mastrangelo, R.; Giorgi, R.; Baglioni, P. Twin-chain polymer networks loaded with nanostructured fluids for the selective removal of a non-original varnish from Picasso's "L'Atelier" at the Peggy Guggenheim Collection, Venice. *Herit. Sci.* **2020**, *8*, 77. [CrossRef]
36. Baglioni, M.; Poggi, G.; Giorgi, R.; Rivella, P.; Ogura, T.; Baglioni, P. Selective removal of over-paintings from "Street Art" using an environmentally friendly nanostructured fluid loaded in highly retentive hydrogels. *J. Colloid Interface Sci.* **2021**, *595*, 187–201. [CrossRef] [PubMed]
37. Lee, C.; Volpi, F.; Fiocco, G.; Weththimuni, M.L.; Licchelli, M.; Malagodi, M. Preliminary cleaning approach with alginate and konjac glucomannan polysaccharide gel for the surfaces of east Asian and western string musical instruments. *Materials* **2022**, *15*, 1100. [CrossRef] [PubMed]
38. Gabriele, F.; Bruno, L.; Casieri, C.; Ranaldi, R.; Rugnini, L.; Spreti, N. Application and monitoring of oxidative alginate–biocide hydrogels for two case studies in "The Sassi and the Park of the Rupestrian Churches of Matera". *Coatings* **2022**, *12*, 462. [CrossRef]
39. Ortega Saez, N.; Arno, R.; Marchetti, A.; Cauberghs, S.; Janssens, K.; Van der Snickt, G.; Al-Emam, E. Towards a novel strategy for soot removal from water-soluble materials: The synergetic effect of hydrogels and cyclomethicone on gelatine emulsion-based photographs. *Herit. Sci.* **2023**, *11*, 78. [CrossRef]
40. Richard, F.; Hermans, J.J.; Angelova, L. Rigid Solvent-Gels in Paper Conservation: A New Approach to Sticky Problems. *J. Pap. Conserv.* **2024**, *25*, 86–106. [CrossRef]
41. Sun, X.; Lu, C.; Liu, Y.; Zhang, W.; Zhang, X. Melt-processed poly (vinyl alcohol) composites filled with microcrystalline cellulose from waste cotton fabrics. *Carbohydr. Polym.* **2014**, *101*, 642–649. [CrossRef]

42. Tretinnikov, O.N.; Zagorskaya, S.A. Determination of the degree of crystallinity of poly (vinyl alcohol) by FTIR spectroscopy. *J. Appl. Spectrosc.* **2012**, *79*, 521–526. [CrossRef]
43. Fei, W.; Wu, Z.; Cheng, H.; Xiong, Y.; Chen, W.; Meng, L. Molecular mobility and morphology change of poly (vinyl alcohol)(PVA) film as induced by plasticizer glycerol. *J. Polym. Sci.* **2023**, *61*, 1959–1970. [CrossRef]

nanomaterials

MDPI

Article

Improving the NO_2 Gas Sensing Performances at Room Temperature Based on TiO_2 NTs/rGO Heterojunction Nanocomposites

Yan Ling [1], Yunjiang Yu [1], Canxin Tian [1,2] and Changwei Zou [1,2,*]

[1] Key Laboratory of Advanced Coating and Surface Engineering, Lingnan Normal University, Zhanjiang 524048, China; lingyan@lingnan.edu.cn (Y.L.); sidui@lingnan.edu.cn (Y.Y.); cxtian@lingnan.edu.cn (C.T.)

[2] Research Center for Engineering Technology in Surface Strengthening of Guangdong Province, Lingnan Normal University, Zhanjiang 524048, China

* Correspondence: zoucw@lingnan.edu.cn; Tel./Fax: +86-759-3183260

Abstract: The development of energy-efficient, sensitive, and reliable gas sensors for monitoring NO_2 concentrations has garnered considerable attention in recent years. In this manuscript, TiO_2 nanotube arrays/reduced graphene oxide nanocomposites with varying rGO contents (TiO_2 NTs/rGO) were synthesized via a two-step method for room temperature NO_2 gas detection. From SEM and TEM images, it is evident that the rGO sheets not only partially surround the TiO_2 nanotubes but also establish interconnection bridges between adjacent nanotubes, which is anticipated to enhance electron–hole separation by facilitating electron transfer. The optimized TiO_2 NTs/rGO sensor demonstrated a sensitive response of 19.1 to 1 ppm of NO_2, a 5.26-fold improvement over the undoped TiO_2 sensor. Additionally, rGO doping significantly enhanced the sensor's response/recovery times, reducing them from 24 s/42 s to 18 s/33 s with just 1 wt.% rGO. These enhancements are attributed to the increased specific surface area, higher concentration of chemisorbed oxygen species, and the formation of p-n heterojunctions between TiO_2 and rGO within the nanocomposites. This study provides valuable insights for the development of TiO_2/graphene-based gas sensors for detecting oxidizing gases at room temperature.

Keywords: TiO_2 nanotubes; reduced graphene oxide; NO_2 gas sensing; nanocomposites; room temperature sensors

1. Introduction

As a harmful air pollutant, nitrogen dioxide (NO_2) has been responsible for significant environmental issues, including acid rain and photochemical smog, which pose serious risks to human health [1,2]. Therefore, the on-line and real-time monitoring of NO_2 leakage is crucial for protecting both public health and environmental safety. The development of energy-efficient, sensitive, and reliable gas sensors for monitoring NO_2 concentrations has obtained considerable attention in recent years [3–5]. Metal oxides are well known for their excellent adsorption capacity, catalytic activity, and thermodynamic stability and are widely utilized in gas sensor applications [6]. Numerous metal oxides have been investigated as potential gas sensors, including ZnO, SnO_2, In_2O_3, WO_3, Fe_2O_3, and TiO_2 [6–12]. However, the performance of NO_2 gas sensors based on metal oxides remains unsatisfactory due to several limitations, such as high operating temperatures, elevated energy consumption, and poor reproducibility [13,14]. Notably, most metal oxide gas sensors require operation at high temperatures ranging from 200 to 400 °C. This not only makes them unsuitable for detecting gases that may contain explosive substances but also poses cost challenges for commercial applications. The prolonged response and recovery times, along with the low sensitivity at room temperature, represent significant bottlenecks for the practical applicability of TiO_2-based gas sensors. To address these issues, it is essential to incorporate or integrate TiO_2 with hybrid nanocomposite materials such as graphene compounds and metal nanoparticles. This approach has been widely regarded as an effective strategy to enhance gas sensing performance.

Due to the introduction of oxygen functional groups and surface defects that serve as active sites for gas adsorption, graphene oxide (GO) and reduced graphene oxide (rGO) are increasingly recognized as suitable materials for room temperature gas sensing [15]. However, due to the limitations in gas sensor fabrication associated with the low defect density of graphene, numerous researchers have sought to develop gas sensors utilizing reduced graphene oxide (rGO), which possesses a higher defect density [16,17]. The development of low cost, transparent, and flexible rGO-based sensors for detecting harmful gases at very low concentrations is significantly important yet still challenging [18]. The TiO_2 nanoparticles/reduced graphene oxide (TiO_2 NPs/rGO) composite demonstrated a significant gas response (~14.9%), which is 4.57 times higher than that of pristine counterparts, along with excellent selectivity, high sensitivity, rapid response and recovery times, as well as remarkable repeatability towards nitrogen dioxide (NO_2) at a concentration of 100 ppm at room temperature [19]. The anti-humidity sensing performance of Pt/GO/TiO_2 is improved by increasing the thickness of the GO interlayer. Remarkably, the diode with a GO areal loading of 0.969 mg cm^{-2} exhibits a response retention rate ($R_{RH95\%}/R_{dry}$) of nearly 100% at 298 K [20]. A variety of high-performance and low-temperature gas sensors based on rGO hybrids have been reported. A straightforward one-pot microwave-assisted hydrothermal method has been employed to synthesize SnO_2/rGO composites, resulting in a significant reduction in the response and recovery times from 39.2/54.7 min to just 6.5/1 min, with an impressive detection limit as low as 50 ppb [21]. Under UV irradiation from an LED, a sensor utilizing graphene/TiO_2 nanoparticles demonstrated a detection limit of approximately 50 ppb for NO_2 at room temperature [22]. NO_2 gas sensors based on a ZnO-rGO hybrid showed improved sensitivity and faster response and recovery times [23]. Furthermore, a low-operating-temperature NO_2 gas sensor based on rGO/SnS_2 has shown remarkable selectivity and reversibility towards NO_2, achieving a low detection limit of 0.6 ppm with a response rate of 9.8% at 80 °C [24]. Moreover, when exposed to 1 ppm NO_2 at room temperature, CuO/rGO hybrids displayed a sensitive response quantified at around 14 [25].

In this manuscript, we fabricated a highly efficient room temperature NO_2 sensor utilizing TiO_2 nanotubes/reduced graphene oxide (TiO_2 NTs/rGO) nanocomposites. In addition, the fabricated gas sensor exhibits a highly sensitive response and excellent selectivity towards NO_2 gas, and the mechanism underlying the gas sensing performance was also investigated.

2. Experimental Details

2.1. Synthesis of Materials

All chemical reagents were of analytical grade (Beijing Chemical Co., Ltd., Beijing, China) and utilized without further purification. TiO_2 nanotube arrays were synthesized through the anodization of titanium foil (99.99%) at a voltage of 45 V for a duration of 2 h. The electrolyte was composed of ethylene glycol (99.99%), 0.3 M ammonium fluoride (NH_4F, 99%), and 2 vol% water (H_2O). Following anodization, the samples were annealed at 400 °C for 2 h, after which they were sonicated for 30 min and subjected to another round of annealing for 2 h. Then, the samples were removed from the bath and allowed to dry at room temperature for 1 h before being further annealed at 600 °C for 2 h in a furnace under air atmosphere. Subsequently, the samples were immersed in a solution containing 3-aminopropyl triethoxysilane and ethanol, followed by refluxing at 80 °C for 2 h. The TiO_2 nanotubes were then thoroughly rinsed with ethanol and deionized water before being dried at room temperature. The reduction of graphene oxide (GO) into reduced graphene oxide (rGO) was accomplished by exposing GO suspension to UV radiation for 1 h. The electrostatic interaction between positively charged nanotubes and negatively charged rGO facilitates the adhesion of graphene derivatives onto the surface of the nanotubes [26]. The theoretical weight percentages of rGO within the TiO_2 NTs/rGO nanocomposites were calculated to be approximately 0.5 wt%, 1 wt%, and 3 wt%. For clarity, these TiO_2 NTs/rGO nanocomposites will henceforth be referred to as containing either 0.5 wt%, 1 wt%, or 3 wt% rGO in subsequent figures. Figure 1a displays a schematic of the synthesis process.

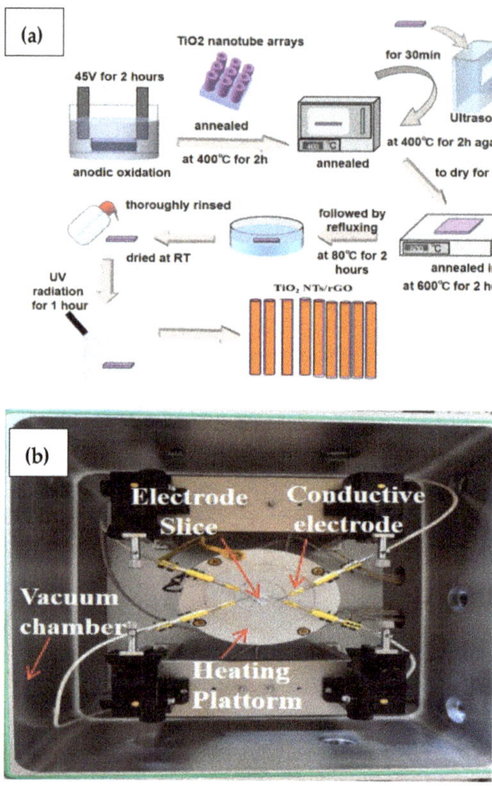

Figure 1. Schematic of synthesis process (**a**) and photograph of NO_2 detection testing system (**b**).

2.2. Characterization

The energy dispersive spectrometry (EDS) and field emission scanning electron microscopy (FESEM) images were acquired using a JEOL JSM-7500F microscope operating at 15 kV (JEOL, Japan). For the X-ray powder diffraction (XRD) analysis, we utilized X-ray diffractometer (XRD; D8, Karlsruhe, Germany) at a scanning rate of 0.02 s^{-1}. The Raman spectroscopy analysis was performed using the RENISHAW INVIA Micro-Raman spectrometer (Renishaw, UK). The X-ray photoelectron spectroscopy (XPS, Kratos XSAM800, Kratos Ltd., Manchester, Britain) was used to examine the chemical bonding states with Mg K_a excitation. Transmission electron microscopy (TEM) and high-resolution transmission electron microscopy (HR-TEM) were conducted on a JEOL JEM-2100F microscope operating at an accelerating voltage of 200 kV. The specific surface area was estimated using the Brunauer–Emmett–Teller (BET) (BET, 3H-2000ps4, China) equation based on nitrogen adsorption isotherms, following prior degassing of the sample under vacuum at 120 °C.

2.3. Fabrication and Measurement of the Gas Sensor

The photograph and structure of the NO_2 detection equipment are presented in Figure 1b [27,28]. The fabrication process can be described as follows: First, a suitable amount of the as-grown TiO_2 NTs/rGO powder was thoroughly mixed with deionized water to create homogeneous slurry. This slurry was then carefully coated onto an alumina tube using a small brush to form a sensing film. A pair of Au electrodes was installed at each end of the tube, with each electrode connected to a pair of Pt wires. After allowing it to dry in air at room temperature, the device underwent annealing at 200 °C for 2 h to eliminate any residual water. Finally, a Ni-Cr alloy coil was inserted into the alumina ceramic tube to serve as a heater. The operating temperature is controlled by adjusting the heating current supplied to the ceramic heater. The response of the sensor is defined as $S = R_g/R_a$, where R_a and R_g represent the electrical resistance of the gas sensor in air and in NO_2, respectively. The response time or recovery time was defined as the time taken for 90% resistance variation.

3. Results and Discussion

3.1. Structural and Morphological Characteristics

Figure 2 presents the XRD patterns of the TiO_2 NTs/rGO nanocomposites synthesized with varying rGO contents. The XRD pattern for undoped TiO_2 nanotubes displays very sharp diffraction peaks, all of which can be confidently assigned to the anatase TiO_2 phase (JCPDS Card No. 21-1272). It is important to note that the incorporation of reduced graphene oxide does not alter the original crystal structure of TiO_2. All diffraction peaks in the nanocomposites are observed at nearly identical 2θ positions when compared to those of undoped TiO_2. However, with an increase in rGO doping, there is a noticeable decrease in the intensity of the diffraction peaks for TiO_2 NTs/rGO nanocomposites, which can be attributed to an excess formation of nucleation centers [29]. Additionally, it is interesting to observe from the XRD pattern of TiO_2 NTs/rGO with an rGO content of 3 wt.% that a weak peak appears between 23° and 26°, which corresponds to the (002) plane of rGO [30,31].

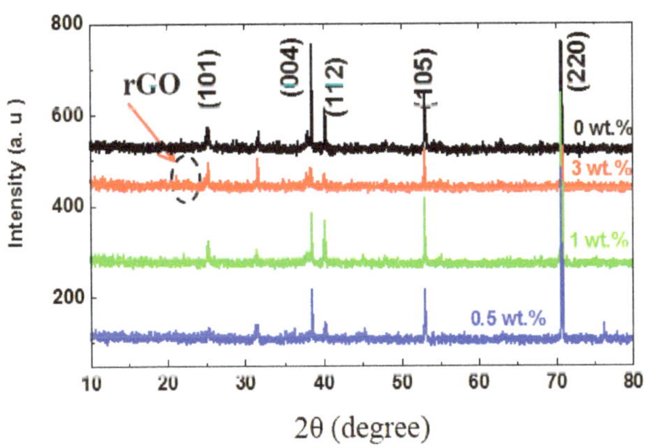

Figure 2. XRD patterns of TiO_2 NTs/rGO nanocomposites grown with different rGO contents.

As illustrated in Figure 3a, the TiO_2 nanotubes are characterized by vertically arranged structures with an average diameter ranging from approximately 80 to 120 nm. For the TiO_2 NTs/rGO nanocomposites with rGO concentrations of 0.5 and 1 wt.%, graphene sheets are not distinctly visible in Fig. 3b and 3c. However, for those with rGO contents of 3 wt.%, it is evident that the rGO sheets not only partially surround the TiO_2 nanotubes but also establish interconnection bridges between adjacent nanotubes (Figure 3d). This connectivity through the rGO layer is anticipated to enhance electron–hole separation by facilitating electron transfer from TiO_2. In addition, electrons traveling along the graphene layer may interact with adsorbed NO_2 molecules as well. To confirm the presence of graphene within these composites, we conducted an EDS analysis on samples with a rGO content of 3 wt.%. The EDX spectrum presented in the inset of Figure 3d reveals peaks corresponding to titanium (Ti), oxygen (O), and carbon (C) elements.

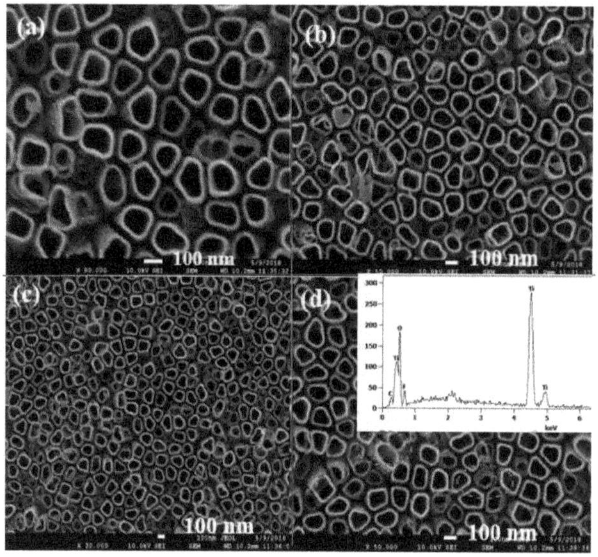

Figure 3. SEM images of TiO_2 NTs/rGO nanocomposites with rGO contents of 0 wt.% (**a**), 0.5 wt.% (**b**), 1 wt.% (**c**), and 3 wt.% (**d**), respectively. Inset of Figure 3d shows the EDX spectrum of TiO_2 NTs/rGO nanocomposites with rGO contents of 3 wt.%. The red circles in Figure 3d illustrate the structure of rGO surrounding the TiO_2 nanotubes.

As can be seen in the TEM images shown in Figure 4a,b, there is no significant difference in diameter and length between TiO_2 nanotubes and TiO_2 NTS/rGO nanocomposites. To gain deeper insights into the nature of the interface, we conducted HR-TEM analysis, with the corresponding image presented in Figure 4d. When compared to the undoped TiO_2 nanotubes shown in Figure 4c, it becomes evident that thin layers are wrapped around the outer surface of the TiO_2 nanotubes, indicating successful modification by rGO. The observed lattice spacing of 0.35 nm for the TiO_2 nanotubes corresponds to the (110) orientation of anatase phase TiO_2. The layer material in contact with TiO_2 is rGO, which forms a close-contact interface with the surface of these nanotubes, which is believed to facilitate electron transmission. However, it is important to note that any observed inhomogeneity within the nanocomposites may arise from limitations associated with chemical processes or reaction times, or the sample preparation method for TEM.

3.2. Raman and XPS Characteristics

The Raman spectra of TiO_2 NTs/rGO nanocomposites, as illustrated in Figure 5, reveal two prominent peaks corresponding to the D and G bands of graphene. The G band provides valuable information regarding the in-plane vibrations of sp^2 bonded carbon atoms [32], while the D band is associated with some sp^3 defects present in rGO [33]. The characteristics of Raman modes of anatase phase appeared at 146, 400, 518, and 636 cm^{-1} for pristine TiO_2, and these peaks are well matched with E_g, B_{1g}, A_{1g} + B_{1g}, and Eg modes of the anatase TiO_2 phase [34]. As depicted in Figure 5, the G band for TiO_2 NTs/rGO composites appears at 1597 cm^{-1}, which is quite close to that of pristine graphene (1580 cm^{-1}). The slight red shift observed in the G band for TiO_2 NTs/rGO nanocomposites compared to GO (1591 cm^{-1}) suggests a restoration of the π-π systems within graphene during the chemical reaction. Additionally, it is widely recognized that the I_D/I_G ratio serves as a valuable indicator of crystal quality within graphite structures. Our calculations indicate that the I_D/I_G ratio for 3 wt.% TiO_2 NTs/rGO (1.26) exceeds that of 0.5 wt.% TiO_2 NTs/rGO (1.14), signifying a stronger D band signal and thus confirming successful incorporation of rGO into these nanocomposites. Additionally, spectral characteristics suggest an increase in defect point density on the resulting rGO sheets, which may serve as ideal adsorption–desorption sites for TiO_2 nanotubes [31].

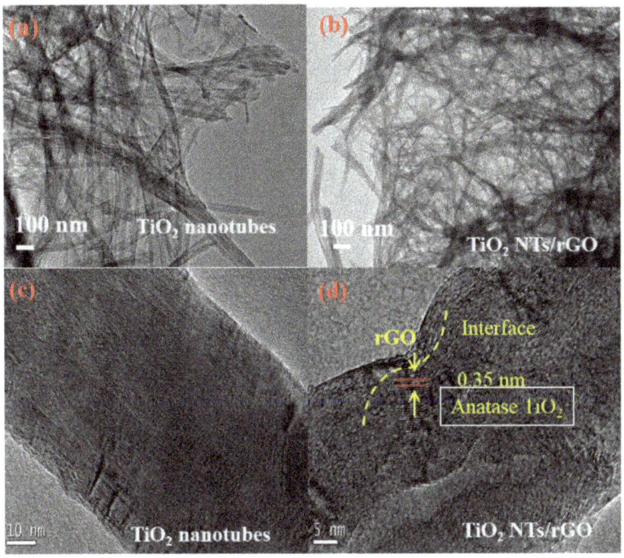

Figure 4. TEM (**a**,**b**) and high-resolution TEM (**c**,**d**) images of TiO_2 nanotubes (**a**,**c**) and TiO_2 NTs/rGO nanocomposites (**b**,**d**) with rGO contents of 1 wt.%.

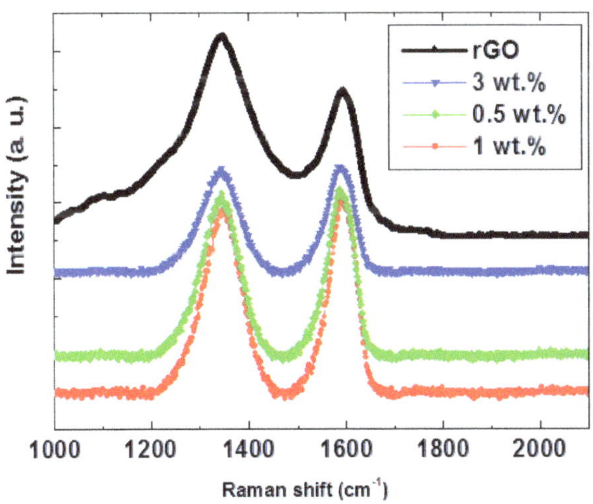

Figure 5. Raman spectra of TiO2 NTs/rGO nanocomposites with different rGO contents.

The chemical states of the elements were carefully analyzed using XPS and are illustrated in Figure 6. The sharp peaks observed in the full scan spectra indicate a clear presence of C, O, and Ti elements for TiO_2 NTs/rGO nanocomposites. The TiO_2 NTs/rGO nanocomposite with a rGO content of 1 wt.% exhibits the best gas sensing performance. Therefore, we selected this sample for XPS characterization. In Figure 6b, the Ti $2p_{3/2}$ and Ti $2p_{1/2}$ peaks are located at binding energies of 458.9 and 464.8 eV, which correspond to the values for Ti^{4+} in TiO_2 [35]. As shown in Figure 6c, the high-resolution C 1s peak was accurately fitted with three distinct components. The binding energy at 284.3 eV, 285.1 eV, and 287.5 eV can be attributed to C-C bonds (sp^2 hybridized carbon) from rGO, C-O-Ti bonds and O-C=O species, respectively [36]. The fitted peak at 285.1 eV corresponding to the C-O-Ti bond further confirms that a chemically bonded heterostructure has indeed formed between TiO_2 NTs and rGO [37]. This heterostructure facilitates close contact between TiO_2 nanotubes and rGO, thereby promoting effective electron transfer. In Figure 6d, we present the high-resolution XPS spectrum for O 1s, which can be resolved into three Gaussian peaks representing different types of oxygen species. The three peaks can be respectively attributed to oxygen vacancy in defective TiO_2, lattice oxygen species (Ti-O bands), and chemisorbed or dissociated oxygen species (C=O) [38]. The increase in oxygen vacancies within defective TiO_2 suggests that there are more active sites available in the TiO_2 NTS/rGO nanocomposites, which is advantageous for gas adsorption and reaction. Additionally, the rise in C=O bonds in these nanocomposites indicates that the oxygen species adsorbed on the surface can participate effectively in surface redox reactions, leading to significant changes in sensor resistance. Therefore, by incorporating rGO, the TiO_2 NTs/rGO nanocomposite demonstrates a remarkable ability to adsorb ionized oxygen, contributing to its potential for high-performance gas sensing applications [39].

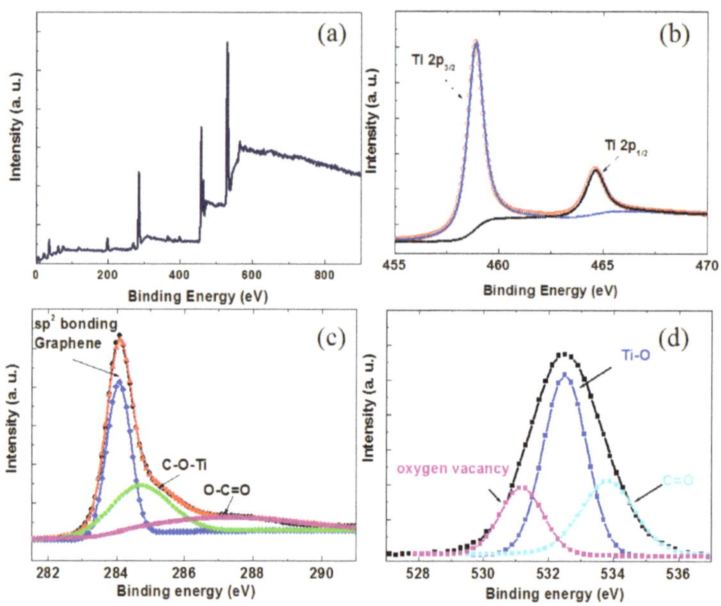

Figure 6. XPS spectra of TiO_2 NTs/rGO nanocomposite with rGO contents of 1 wt.%. (**a**) Full scan. (**b**) Ti 2p. (**c**) C 1s. (**d**) O 1s.

3.3. Gas Sensing Properties

The relationship between the response and NO_2 concentration for the sensor based on TiO_2 NTs/rGO nanocomposites at room temperature is illustrated in Figure 7. It can be observed that the response increases with NO_2 concentration, ranging from 1 ppm to 50 ppm. Notably, the TiO_2 NTs/rGO nanocomposite with contents of 1.0 wt.% exhibited the highest response value, reaching an impressive 138 at a NO_2 concentration of 50 ppm. The limit of detection was determined to be 1 ppm, and it is worth mentioning that the sensor demonstrated a remarkable response of 19.1 for just 1 ppm NO_2 when utilizing the TiO_2 NTs/rGO composite with rGO contents of 1.0 wt.%.

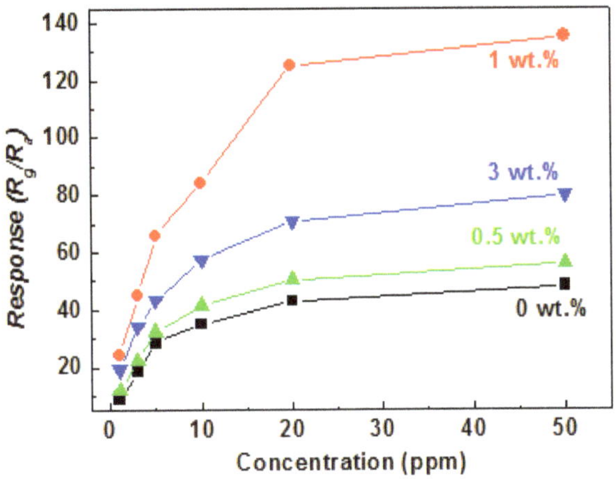

Figure 7. The response value of the sensor based on TiO_2 NTs/rGO composites vs. NO_2 concentration at room temperature.

The response and recovery curves of the sensor to 20 ppm NO_2 are presented in Figure 8. The calculated response/recovery times are as follows: 24 s/42 s for TiO_2 nanotubes, 23 s/34 s for TiO_2 NTs/rGO (0.5 wt.%), 18 s/33 s for TiO_2 NTs/rGO (1 wt.%), and 20 s/35 s for TiO_2 NTs/rGO (3 wt.%). It is evident that the sensors based on the 1 wt.% rGO doped TiO_2 NTs/rGO exhibit a significantly faster response time. However, it is worth noting that the response time tends to increase with higher doping amounts of rGO. The relative filling and partial binding of TiO_2 nanotubes due to rGO doping positively influence gas permeation into the sensing layer as well as electron transfer from TiO_2 to graphene, which may contribute to a quicker reduction in resistance. Nevertheless, an increase in rGO doping can lead to agglomeration, which reduces the active sites available for gas molecules and consequently diminishes resistance charge. The room temperature NO_2 gas sensing performances of TiO_2 NTs/rGO were compared with previous metal-semiconductor oxide/rGO nanocomposites. In Table 1, we can see that in low concentration detection and response/recovery time, TiO_2 NTs/rGO exhibited better gas sensing performance than most of the previously reported composites.

Table 1. Comparison of the RT NO_2 gas sensing performances of our device with those reported in previous literature.

Ref.	Device Structure	Target Gas	Working Temperature (°C)	Response (%)/ppm	Response/Recovery Time (s)
[3]	rGO/CeO_2	NO_2	RT	8.2/25	180/260
[4]	In_2O_3/rGO	NO_2	RT	8.25/30	165/235
[5]	Graphene/ZnO	NO_2	300	9.5/50	145/248
[40]	TiO_2 NPs/rGO	NO_2	RT	14.9/100	124/182
This work	TiO_2 NTs/rGO	NO_2	RT	19.1/1	18 s/33 s

The selectivity of gas sensors is another crucial parameter for real-time applications. As illustrated in Figure 9, the sensor utilizing TiO_2 NTs/rGO composites (1 wt. %) demonstrated an impressive response to NO_2, with its value being at least four times greater than that observed for other test gases. This behavior regarding CO, H_2S, and H_2 can be attributed to the low operating temperature and relatively low concentration of the detected gases [41]. Based on these findings, it is evident that the sensor employing TiO_2 NTS/rGO nanocomposites exhibits superior performance in detecting low concentrations of NO_2 at lower temperatures compared to other detection gases.

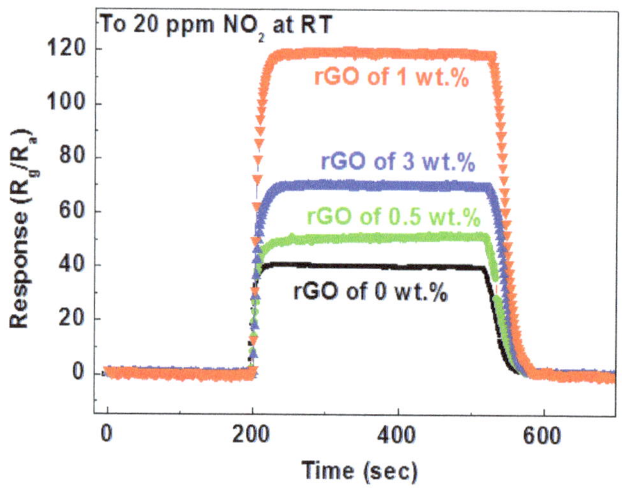

Figure 8. The dynamic response transients of the sensor based on TiO_2 NTs/rGO nanocomposites to 20 ppm NO_2 at room temperature.

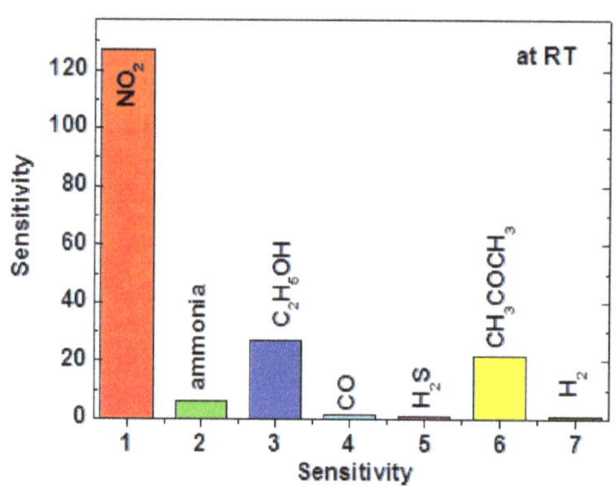

Figure 9. Response of the sensor based on TiO_2 NTs/rGO nanocomposites to 50 ppm of C_2H_5OH, CH_3OH, H_2, NH_3, H_2S, and NO_2 at room temperature.

The repeatability and long-term stability of the sensor were investigated, providing crucial parameters for practical applications. The long-term stability of the sensors based on TiO_2 nanotubes (NTs) and TiO_2 NTs/reduced graphene oxide (rGO) nanocomposites in response to 1 ppm NO_2 at room temperature is illustrated in Figure 10. The sensor utilizing TiO_2 NTs/rGO (1 wt.%) exhibited a stable response of 17.2 (17.2 $\pm$ 0.3) for 1 ppm NO_2 at room temperature, demonstrating excellent long-term stability.

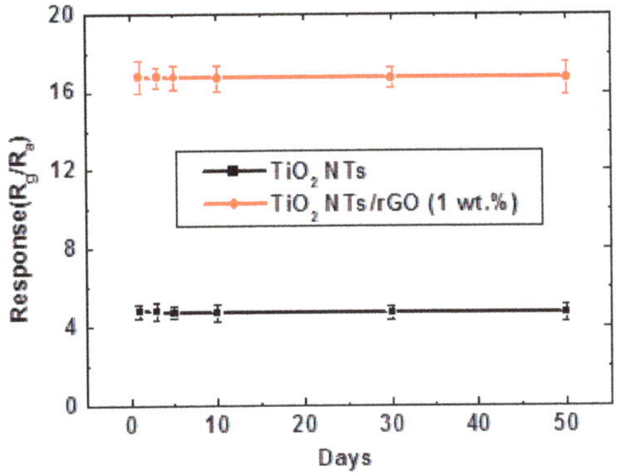

Figure 10. Long-term stability of the sensor based on TiO_2 NTs and TiO_2 NTs/rGO nanocomposites to 1 ppm of NO_2 at room temperature.

3.4. Gas Sensing Mechanism of TiO_2 NTs/rGO Nanocomposites

The sensing principle of resistance is fundamentally based on the changes in sensor resistance, which occurs due to variations in charge carriers and is closely linked to the amount of chemically adsorbed substances present on the surface of metal oxides. In the case of *n-type* TiO_2, O_2 molecules from the atmosphere are adsorbed onto its surface and subsequently transform into O^-, O_2^-, and O^{2-} by capturing electrons from TiO_2. However, when NO_2 is introduced, its chemical adsorption further reduces conductivity through an electron capture effect. We have observed that incorporating rGO can significantly enhance the conductivity of these sensors, thereby improving their gas sensing performance at room temperature. As illustrated by the I-V curves shown in Figure 11, it becomes evident that the TiO_2 NTs/rGO nanocomposite exhibits much lower resistance compared to undoped TiO_2 nanotubes, which indicates a notable improvement in charge transfer performance. Moreover, with an increase in active sites such as vacancies, defects, and oxygen functional groups, we expect a significant improvement in both the gas adsorption and diffusion rates of NO_2 molecules on these active surfaces.

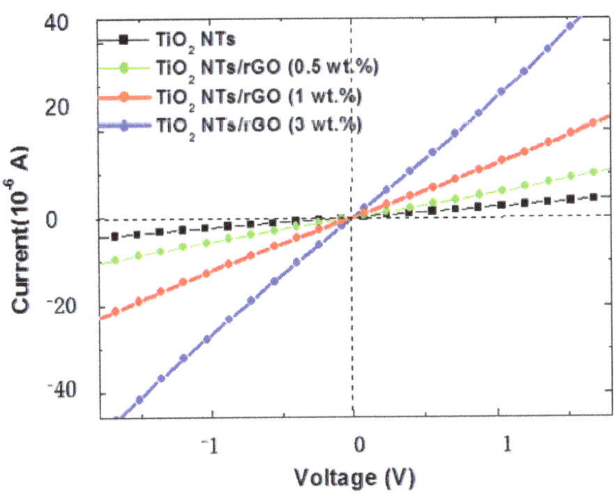

Figure 11. I-V curves of the sensor based on TiO_2 NTs/rGO nanocomposites with different rGO contents.

Another mechanism contributing to the enhancement of the sensor can be attributed to the formation of a *p-n* heterojunction between *n-type* TiO_2 and *p-type* rGO. The close contact between these two distinct semiconductor materials allows for the alignment of their Fermi energy levels at the interface, which typically results in charge transfer and the establishment of a charge depletion layer. Figure 12a illustrates the energy band diagrams for TiO_2 nanotubes, rGO, and their corresponding TiO_2 NTs/rGO nanocomposites. The work function for TiO_2 NTs is approximately 5.1 eV, while that of rGO is around 4.42 eV. Upon forming the TiO_2/rGO heterojunction, an electron accumulation region develops on the surface of TiO_2 NTs. This charge transfer creates a potential barrier at the heterojunction, leading to bending in both vacuum energy levels and energy bands. The initial electron transfer from TiO_2 to graphene generates a surface depletion region on TiO_2, resulting in increased resistance. In ambient air conditions, where no barriers exist between TiO_2 NTs and graphene, electrons can flow freely from TiO_2 NTs to graphene. However, upon exposure to NO_2 gas, there is an increase in potential barrier height at the interfaces between *n-type* TiO_2 NTs and *p-type* rGO, which makes electron transfer from *p-type* graphene to *n-type* TiO_2 NTs more challenging and ultimately leads to an increase in sensor resistance (Figure 12b).

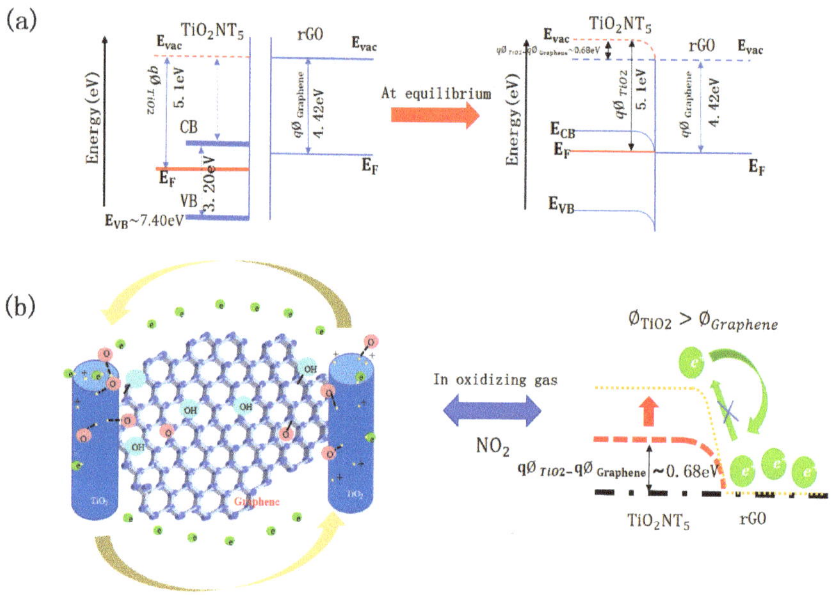

Figure 12. Energy band diagrams for TiO_2 NTs, rGO and TiO_2 NTs/rGO heterostructure, where E_{VB}, E_F, E_{CB}, E_{vac} represent valence band, Fermi level, conduction band, and vacuum level, respectively. (**a**) The energy band diagrams for TiO_2 nanotubes, rGO, and their corresponding TiO_2 NTs/rGO nanocomposites. (**b**) Schematic illustration of electron transfer and sensing mechanism of TiO_2 NTs/rGO nanocomposites.

4. Conclusions

In summary, TiO_2 nanotube arrays modified by rGO nanosheets were successfully synthesized using a straightforward two-step method. The SEM images reveal that the rGO layer not only partially enveloped the TiO_2 nanotubes but also established a close connection between adjacent nanotubes, thereby enhancing the efficiency of electron–hole separation. The gas sensing properties of the TiO_2 NTs/rGO nanocomposites demonstrated high sensitivity to NO_2 at low operating temperatures, characterized by an enhanced response, relatively short recovery time, and excellent selectivity. The improved gas sensing performance can be attributed to the formation of a *p-n* heterojunction between *n-type* TiO_2 and *p-type* rGO. Furthermore, due to the presence of rGO sheets, the rate of gas adsorption and diffusion on the active surface of TiO_2 NTs/rGO nanocomposites was significantly facilitated by an increased number of active sites.

Author Contributions: Methodology, Y.Y. and C.T.; Investigation, Y.L., C.T. and C.Z.; Data curation, Y.L. and C.T.; Writing—original draft, Y.L., Y.Y. and C.Z. All authors have read and agreed to the published version of the manuscript.

Funding: This research was funded by the Innovation team project of Lingnan Normal University (LT2203), Science and Technology Project of Zhanjiang (2019A03009, 2022A01001, 2023A21510), and Natural Science Foundation of Guangdong Province (2022A1515011137, 2023A1515011796, 2023A1515012923).

Data Availability Statement: Data are contained within the article.

Conflicts of Interest: The authors declare no conflict of interest.

References

1. Deng, S.; Tjoa, V.; Fan, H.M.; Tan, H.R.; Sayle, D.C.; Olivo, M.; Mhaisalkar, S.; Wei, J.; Sow, C.H. Reduced graphene oxide conjugated Cu2O nanowire mesocrystals for high-performance NO2 gas sensor. *J. Am. Chem. Soc.* **2012**, *134*, 4905–4917. [CrossRef] [PubMed]
2. Turlybekuly, A.; Sarsembina, M.; Mentbayeva, A.; Bakenov, Z.; Soltabayev, B. CuO/TiO_2 heterostructure-based sensors for conductometric NO_2 and N_2O gas detection at room temperature. *Sens. Actuators B* **2023**, *397*, 134635. [CrossRef]
3. Hu, J.; Zou, C.; Su, Y.; Li, M.; Ye, X.; Cai, B.; Kong, E.S.; Yang, Z.; Zhang, Y. Light-assisted recovery for a highly-sensitive NO_2 sensor based on RGO-CeO_2 hybrids. *Sens. Actuators B Chem.* **2018**, *270*, 119–129. [CrossRef]
4. Gu, F.; Nie, R.; Han, D.; Wang, Z. In_2O_3–graphene nanocomposite based gas sensor for selective detection of NO_2 at room temperature. *Sens. Actuators B Chem.* **2015**, *219*, 94–99. [CrossRef]
5. Kim, H.W.; Kwon, Y.J.; Mirzaei, A.; Kang, S.Y.; Choi, M.S.; Bang, J.H.; Kim, S.S. Synthesis of zinc oxide semiconductors-graphene nanocomposites by microwave irradiation for application to gas sensors. *Sens. Actuators B Chem.* **2017**, *249*, 590–601. [CrossRef]
6. Young, S.-J.; Chu, Y.-L. Hydrothermal Synthesis and Improved CH_3OH-Sensing Performance of ZnO Nanorods with Adsorbed Au NPs. *IEEE Trans. Electron Devices* **2021**, *68*, 1886–1891. [CrossRef]
7. Chu, Y.-L.; Young, S.-J.; Ji, L.-W.; Chu, T.-T.; Lam, K.-T.; Hsiao, Y.-J.; Tang, I.-T.; Kuo, T.-H. Characteristics of Gas Sensors Based on Co-Doped ZnO Nanorod Arrays. *J. Electrochem. Soc.* **2020**, *167*, 117503. [CrossRef]
8. Zhang, D.H.; Liu, Z.Q.; Li, C.; Tang, T.; Liu, X.L.; Han, S.; Lei, B.; Zhou, C.W. Detection of NO_2 down to ppb levels using individual and multiple In_2O_3 nanowire devices. *Nano Lett.* **2004**, *4*, 1919–1924. [CrossRef]
9. Tsai, Y.T.; Chang, S.J.; Ji, L.W.; Hsiao, Y.J.; Tang, I.T.; Lu, H.Y.; Chu, Y.L. High Sensitivity of NO Gas Sensors Based on Novel Ag-Doped ZnO Nanoflowers Enhanced with a UV Light-Emitting Diode. *ACS Omega* **2018**, *3*, 13798–13807. [CrossRef]
10. Wetchakun, K.; Samerjai, T.; Tamaekong, N.; Liewhiran, C.; Siriwong, C.; Kruefu, V.; Wisitsoraat, A.; Tuantranont, A.; Phanichphant, S. Semiconducting metal oxides as sensors for environmentally hazardous gases. *Sens. Actuators B-Chem.* **2011**, *160*, 580–591. [CrossRef]
11. Young, S.-J.; Chu, Y.-L. Platinum Nanoparticle-Decorated ZnO Nanorods Improved the Performance of Methanol Gas Sensor. *J. Electrochem. Soc.* **2020**, *167*, 147508. [CrossRef]
12. Qin, Y.; Zhang, X.; Wang, Y. Remarkable improvement of $W_{18}O_{49}$/TiO_2 heteronanowires in ambient temperature-responsive NO_2-sensing abilities and its unexpected n-p transition phenomenon. *Sens. Actuators B* **2017**, *240*, 477–486. [CrossRef]
13. Korotcenkov, G. The role of morphology and crystallographic structure of metal oxides in response of conductometric-type gas sensors. *Mater. Sci. Eng. R* **2008**, *61*, 1–39. [CrossRef]
14. Sun, F.Q.; Cai, W.P.; Li, Y.; Jia, L.C.; Lu, F. Direct growth of mono- and multilayer nanostructured porous films on curved surfaces and their application as gas sensors. *Adv. Mater.* **2005**, *17*, 2872–2877. [CrossRef]
15. Yuan, W.J.; Liu, A.R.; Huang, L.; Li, C.; Shi, G.Q. High-performance NO_2 sensors based on chemically modified graphene. *Adv. Mater.* **2013**, *25*, 766–771. [CrossRef]
16. Jung, M.W.; Kang, S.M.; Nam, K.H.; An, K.S.; Ku, B.C. Highly transparent and flexible NO_2 gas sensor film based on MoS_2/rGO composites using soft lithographic patterning. *Appl. Suf. Sci.* **2018**, *456*, 7–12. [CrossRef]
17. Liu, X.; Sun, J.B.; Zhang, X.T. Novel 3D graphene aerogel-ZnO composites as efficient detection for NO_2 at room temperature. *Sens. Actuators B* **2015**, *211*, 220–226. [CrossRef]
18. Varghese, S.S.; Lonkar, S.; Singh, K.K.; Swaminathan, S.; Abdala, A. Recent advances in graphene based gas sensors. *Sens. Actuat. B Chem.* **2015**, *218*, 160–183. [CrossRef]
19. Murali, G.; Reddeppa, M.; Seshendra Reddy, C.; Park, S.; Chandrakalavathi, T.; Kim, M.D.; In, I. Enhancing the charge carrier separation and transport via nitrogen-doped graphene quantum dot-TiO_2 nanoplate hybrid structure for an efficient NO gas sensor. *ACS Appl. Mater. Interfaces* **2020**, *12*, 13428–13436. [CrossRef]
20. Li, R.; Xu, Z.; Mao, B.; Xie, G.; Wang, Y.; Liu, Z.; Fang, H. Pt/GO/TiO_2 room-temperature hydrogen sensing Schottky diode: High resistance to humidity interference endowed by the graphene oxide interlayer. *Sens. Actuators B Chem.* **2024**, *408*, 135569. [CrossRef]
21. Wang, Z.; Jia, Z.; Li, Q.; Zhang, X.; Sun, W.; Sun, J.; Liu, B.; Ha, B. The enhanced NO_2 sensing properties of SnO_2 nanoparticles/reduced graphene oxide composite. *J. Colloid Interface Sci.* **2019**, *537*, 228–237. [CrossRef] [PubMed]

22. Giampiccolo, A.; Tobaldi, D.M.; Leonardi, S.G.; Murdoch, B.J.; Seabra, M.P.; Ansell, M.P.; Neri, G.; Ball, R.J. Sol gel graphene/TiO_2 nanoparticles for the photocatalytic-assisted sensing and abatement of NO_2. *Appl. Catal. B Environ.* **2019**, *243*, 183–194. [CrossRef]
23. Liu, S.; Yu, B.; Zhang, H.; Fei, T.; Zhang, T. Enhancing NO_2 gas sensing performances at room temperature based on reduced graphene oxide-ZnO nanoparticles hybrids. *Sens. Actuators B Chem.* **2014**, *202*, 272–278. [CrossRef]
24. Shafiei, M.; Bradford, J.; Khan, H.; Piloto, C.; Wlodarski, W.; Li, Y.; Motta, N. Low-operating temperature NO_2 gas sensors based on hybrid two-dimensional SnS_2-reduced graphene oxide. *Appl. Surf. Sci.* **2018**, *462*, 330–336. [CrossRef]
25. Li, Z.; Liu, Y.; Guo, D.; Guo, J.; Su, Y. Room-temperature synthesis of CuO/reduced graphene oxide nanohybrids for high-performance NO_2 gas sensor. *Sens. Actuators B Chem.* **2018**, *271*, 306–310. [CrossRef]
26. Rambabu, Y.; Kumar, U.; Singhal, N.; Kaushal, M.; Jaiswal, M.; Jain, S.L.; Roy, S.C. Photocatalytic reduction of carbon dioxide using graphene oxide wrapped TiO_2 nanotubes. *Appl. Surf. Sci.* **2019**, *485*, 48–55. [CrossRef]
27. Wang, J.; Ke, L.; Wu, J.; Liang, F.; Xiang, Y. Research and Development of Online Monitoring Protection Sensors for Paper Books Based on TiO_2 NT/MoS_2. *Coatings* **2024**, *14*, 552. [CrossRef]
28. Wang, J.; Wang, Q.; He, S.; Chen, Z.; Qiu, W.; Yu, Y. Development and Application of a Nano-Gas Sensor for Monitoring and Preservation of Ancient Books in the Library. *Coatings* **2024**, *14*, 553. [CrossRef]
29. Lv, K.; Fang, S.; Si, L.; Xia, Y.; Ho, W.; Li, M. Fabrication of TiO_2 nanorod assembly grafted rGO (rGO@TiO_2-NR) hybridized flake-like photocatalyst. *Appl. Surf. Sci.* **2017**, *391*, 218–227. [CrossRef]
30. Zhang, X.Y.; Li, H.P.; Cui, X.L.; Lin, Y.H. Graphene/TiO_2 nanocomposites: Synthesis, characterization and application in hydrogen evolution from water photocatalytic splitting. *J. Mater. Chem.* **2010**, *20*, 2801–2806. [CrossRef]
31. Pimenta, M.A.; Dresselhaus, G.; Dresselhaus, M.S.; Cancado, L.G.; Jorio, A.; Saito, R. Studying disordering graphite-based systems by Raman spectroscopy. *Phys. Chem. Chem. Phys.* **2007**, *9*, 1276–1290. [CrossRef] [PubMed]
32. Zhang, W.; Cui, J.; Tao, C.; Wu, Y.; Li, Z.; Ma, L. A strategy for producing pure single-layer graphene sheets based on a confined self-assembly approach. *Angew. Chem. Int. Ed.* **2009**, *48*, 5848–5864. [CrossRef] [PubMed]
33. Moussa, S.; Panchakarla, L.S.; Ho, M.Q.; El-Shall, M.S. Graphene supported iron-based nanoparticles for catalytic production of liquid hydrocarbons from synthesis gas. *ACS Catal.* **2014**, *4*, 535–545. [CrossRef]
34. Saha, R.; Thukral, A.; Pal, A.; Saini, P. Sunlight responsive superhydrophilic rGO-TiO_2 nanocomposite coatings for photovoltaic applications. *Diam. Relat. Mater.* **2024**, *149*, 111603. [CrossRef]
35. Lin, Z.; Waller, G.; Liu, Y.; Liu, M.; Wong, C.P. Facile synthesis of nitrogen-doped graphene via pyrolysis of graphene oxide and urea, and its electrocatalytic activity toward the oxygen-reduction reaction. *Adv. Energy Mater.* **2012**, *2*, 884. [CrossRef]
36. Wang, C.; Xue, J.; Qiang, L.; Tang, J. A facile and efficient method to directly synthesize TiO_2/rGO with enhanced photocatalytic performance. *Superlattices Microstruct.* **2018**, *121*, 1–8.
37. Dong, F.; Guo, S.; Wang, H.; Li, X.; Wu, Z. Enhancement of the visible light photocatalytic activity of C-doped TiO_2 nanomaterials prepared by a green synthetic approach. *J. Phys. Chem. C* **2011**, *115*, 13285–13292. [CrossRef]
38. Perera, S.; Mariano, R.; Vu, K.; Nour, N.; Seitz, O.; Chabal, Y.; Balkus, K. Hydrothermal synthesis of graphene-TiO_2 nanotube composites with enhanced photocatalytic activity. *ACS Catal.* **2012**, *2*, 949–956. [CrossRef]
39. Niu, X.; Yan, W.; Zhao, H.; Yang, J. Synthesis of Nb doped TiO_2 nanotube/reduced graphene oxide heterostructure photocatalyst with high visible light photocatalytic activity. *Appl. Surf. Sci.* **2018**, *440*, 804–813. [CrossRef]
40. Harathi, N.; Bollu, M.; Pasupuleti, K.S.; Tauanov, Z.; Peta, K.R.; Kim, M.-D.; Reddeppa, M.; Sarkar, A.; Rao, V.N. PrGO decorated TiO_2 nanoplates hybrid nanocomposite for augmented NO_2 gas detection with faster gas kinetics under UV light irradiation. *Sens. Actuators B Chem.* **2022**, *358*, 131503. [CrossRef]
41. Maziarz, W. TiO_2/SnO_2 and TiO_2/CuO thin film nano-heterostructures as gas sensors. *Appl. Sur. Sci.* **2019**, *480*, 361–370. [CrossRef]

nanomaterials

MDPI

Article

Anti-Sintering Behavior of GYYSZ, Thermophysical Properties, and Thermal Shock Behavior of Thermal Barrier Coating with YSZ/Composite/GYYSZ System by Atmospheric Plasma Spraying

Chunxia Jiang [1,2], Rongbin Li [1,*], Feng He [1], Zhijun Cheng [1], Wenge Li [2] and Yuantao Zhao [2]

[1] School of Materials Science and Engineering, Shanghai Dianji University, Shanghai 201306, China; jiangcx@sdju.edu.cn (C.J.); haggsicff@163.com (F.H.); 181002500112@sdju.edu.cn (Z.C.)

[2] Merchant Marine College, Shanghai Maritime University, Shanghai 201306, China; wgli@shmtu.edu.cn (W.L.); zhaoyt@shmtu.edu.cn (Y.Z.)

* Correspondence: lirb@sdju.edu.cn; Tel.: +86-13817841366

Abstract: In this study, Gd_2O_3 and Yb_2O_3 co-doped YSZ (GYYSZ) ceramic coatings were prepared via atmospheric plasma spraying (APS). The GYYSZ ceramic coatings were subjected to heat treatment at different temperatures for 5 h to analyze their high-temperature phase stability and sintering resistance. The thermophysical properties of GYYSZ, YSZ, and composite coatings were compared. Three types of thermal barrier coatings (TBCs) were designed: GYYSZ (TBC-1), YSZ/GYYSZ (TBC-2), and YSZ/Composite/GYYSZ (TBC-3). The failure mechanisms of these three TBCs were investigated. The results indicate that both the powder and the sprayed GYYSZ primarily maintain a homogeneous cubic phase c-ZrO_2, remaining stable at 1500 °C after annealing. The sintering and densification of the coatings are influenced by the annealing temperature; higher temperatures lead to faster sintering rates. At 1500 °C, the grain size and porosity of GYYSZ are 4.66 μm and 9.9%, respectively. At 1000 °C, the thermal conductivity of GYYSZ is 1.35 W m^{-1} K^{-1}, which is 44% lower than that of YSZ. The thermal conductivity of the composite material remains between 1.79 W·m^{-1} K^{-1} and 1.99 W·m^{-1} K^{-1} from room temperature to 1000 °C, positioned between GYYSZ and YSZ. In the TBC thermal shock water quenching experiment, TBC-3 demonstrated an exceptionally long thermal shock lifetime of 246.3 cycles, which is 5.8 times that of TBC-1 and 1.8 times that of TBC-2. The gradient coating structure effectively reduces the thermal mismatch stress between layers, while the dense surface microcracks provide a certain toughening effect. Failure analysis of the TBC reveals that TBC-3 exhibits a mixed failure mode characterized by both spallation and localized peeling. The ultimate failure was attributed to the propagation of transverse cracks during the final stage of water quenching, which led to the eventual spallation of the ceramic blocks.

Keywords: thermal barrier coating; GYYSZ; atmospheric plasma spraying; phase stability; anti-sintering behavior; composite gradient coating; thermal shock resistance

1. Introduction

Gas turbines are the core power equipment in thermal energy systems, widely used in electricity generation, marine, locomotive, and transportation sectors [1]. Currently, the inlet gas temperature for G/H class gas turbine stage 1 blades has reached 1370−1450 °C, while J class turbines experience inlet temperatures as high as 1600 °C [2]. TBC technology involves depositing materials with low thermal conductivity and phase stability at a high temperature onto the surface of high-temperature alloys, protecting the substrate from thermal damage. YSZ is widely utilized in gas turbines and aero engines due to its excellent thermal cycling performance, high melting point, and low thermal conductivity [3,4]. However, with industrial advancements, the shortcomings of YSZ have become apparent;

it easily undergoes phase transformations, severe sintering, and oxidation when operating above 1200 °C. This not only diminishes the heat-insulating property of the coating but also alters internal stresses, leading to eventual spallation [5–8]. To achieve higher thermal efficiency in gas turbines, traditional YSZ TBCs are increasingly inadequate for elevated temperature demands, necessitating the development of new thermal barrier coatings with superior performance for more complex environments.

Doping YSZ with rare earth oxides can significantly enhance its thermophysical properties [9,10]. Research has found that doping with materials such as CeO_2, Yb_2O_3, Er_2O_3, and Gd_2O_3 could reduce the YSZ coating's thermal conductivity effectively. The reason for this is the differences in ionic radii between the dopants and Zr^{4+}, which generate a greater number of oxygen vacancies and an increased concentration of lattice defects. Such changes enhance phonon scattering, thereby lowering thermal conductivity [11,12]. Partial substitution of Zr^{4+} with rare earth or alkaline earth oxides forms a substituted solid solution, which helps stabilize the tetragonal phase of ZrO_2, allowing the material to maintain phase stability [13,14]. Because of the variations in mass and ionic radius between rare earth oxides and Zr^{4+} or Y^{3+}, they can create a multi-point dislocation-complexing effect. This not only enhances the mechanical properties, sintering resistance, and stability of the materials but also improves the thermal cycling lifetimes of TBCs [15,16]. Chen [17] and Lei [18] prepared Yb_2O_3-Gd_2O_3 co-doped YSZ coatings of varying thicknesses using the APS method. Incorporating rare earth oxides can form an impermeable reaction layer, enhancing the thermal stability and corrosion resistance of YSZ, with multiple doping effects proving to be more effective [19–21]. Chen's investigation of LGYYSZ revealed that the increased number of oxygen vacancies and defects, along with the mismatch in mass and size between these defects and the matrix lattice, diminished the average phonon mean free path. Consequently, the thermal conductivity of LGYYSZ ceramics at 1000 °C was recorded at 1.21 $W{\cdot}m^{-1}{\cdot}K^{-1}$, which is roughly 40% lower than that of YSZ. YbGdZrO ceramics sintered for 400 h at 1723 K exhibited no new phase formation and demonstrated excellent thermal stability [15]. The arithmetic average of the thermal diffusivity for YbGdZrO bulk at 1273 K was found to be 0.385 mm^2/s, which is comparable to that of $Gd_2Zr_2O_7$ ceramics (0.388 mm^2/s at 1273 K) [22].

Nevertheless, these materials frequently demonstrate relatively low thermal expansion coefficients and inadequate mechanical properties. To address these shortcomings, extensive research and improvements have been made regarding TBCs in recent years. In addition to the typical bilayer thermal barrier coating structure, studies have also explored bilayer ceramic structures and functionally graded thermal barrier coatings. Bilayer ceramic coatings not only enhance the working temperature of the coatings but also extend their service life. Wang [23] used APS to prepare YSZ bilayer coatings and $La_2Zr_2O_7$/8YSZ bilayer ceramic coatings. Experimental results showed that in the 1000 °C water quenching thermal cycling test, the bilayer structure failed at 46 times, while the bilayer ceramic thermal barrier coating reached 158 cycles. In the 1200 °C water quenching thermal cycling test, the thermal shock lifetimes were 7 and 43 cycles, respectively, significantly enhancing thermal shock performance. Xu [24] employed the EB-PVD to fabricate $La_2Zr_2O_7$/8YSZ bilayer ceramic coatings, investigating their cyclic oxidation behavior. To further improve the interfacial bonding among the components, functionally graded thermal barrier coatings (FGTBCs) were designed. Numerical simulations have indicated that FGTBCs exhibit significantly enhanced thermal shock lifetimes [25,26]. Portinha [27] designed a porosity gradient structure, achieving over 100 cycles in thermal shock tests. Chen [28] conducted performance studies on $La_2Zr_2O_7$/8YSZ, finding that the gradient coatings significantly outperformed traditional bilayer structures in thermal shock resistance. Additionally, Chen [29] developed an LMA/YSZ functionally graded TBC system. Experimental results indicated that at 1350 °C, the gradient coating exhibited a more durable thermal shock lifetime. Although the introduction of gradient structures has improved the thermal shock resistance and extended the service life of TBCs, the complexity of the gradient structure fabrication process poses challenges in accurately controlling the gradient profiles.

Thus, there remains significant room for development in the research of gradient structure thermal barrier coatings.

GYYSZ is a promising new TBC material. Research on GYYSZ gradient thermal barrier coatings is still limited. This study employed APS technology to fabricate GYYSZ ceramic coatings, investigating their high-temperature phase stability and sintering resistance at various annealing temperatures. Bulk ceramics were produced through hot pressing, and the thermophysical properties, including the thermal conductivity and thermal expansion coefficients of GYYSZ, YSZ, and composite materials, were tested and evaluated. Three TBC structures were designed and fabricated: bilayer structures, bilayer ceramic structures, and functionally graded structures. The thermal shock resistance of these three TBCs was assessed through water quenching thermal shock experiments, exploring the failure mechanisms of each structure from the perspective of crack development.

2. Materials and Methods

The thermal barrier coating substrate material selected for this experiment is the 718 nickel alloy. Its dimensions are ϕ 25 mm × 5 mm. The sprayed powder has a micrometer-scale particle size. The bonding layer of NiCrAlY uses commercial powder Amdry 962 from Oerlikon Metco, Vista, NSW, Australia, with the composition of Ni-22Cr-10Al-1.0Y. The ceramic layers consist of YSZ (ZrO_2-8Y_2O_3, Metco 204NS, Oerlikon Metco) and GYYSZ (ZrO_2-9.5Y_2O_3-5.6Yb_2O_3-5.2Gd_2O_3, Metco 206A, Oerlikon Metco). The composite coating is prepared by mixing YSZ and GYYSZ powders in a mass ratio of 1:1, with the composite powder created through ball milling prior to spraying. To investigate the high-temperature phase stability and sintering resistance of GYYSZ, coatings were sprayed onto 304 stainless steel and subjected to multiple water quenching processes to obtain independent ceramic layers. For assessing the thermophysical properties of the coating materials, ceramic bulk samples of YSZ, GYYSZ, and composite structures were fabricated using ball milling and hot pressing, with dimensions of ϕ 12.7 × 1 mm and ϕ 6 × 25 mm. The coatings were prepared using the Multicoat System plasma spraying system from Oerlikon Metco. Due to the varying characteristics of different powders, experimental parameters listed in Table 1 were selected for this study after consulting the relevant literature and conducting multiple experimental adjustments [30,31]. The designs of the three TBC structures are illustrated in Figure 1. The number of samples sprayed using the same spraying process is 9.

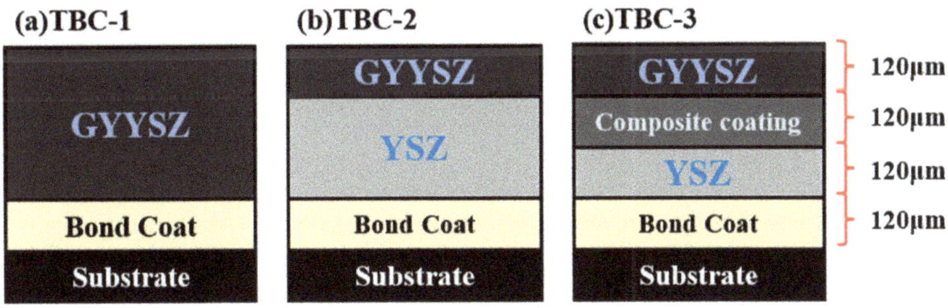

Figure 1. The schematic diagrams of the TBC structures: (**a**) bilayer; (**b**) bilayer ceramic; (**c**) functionally graded ceramic.

Table 1. Spray parameters for the NiCrAlY bond coat and the TBCs.

Parameters	NiCrAlY Bond Coat	YSZ Coat	GYYSZ Coat	Composite Coat
Current (A)	600	550	600	600
Primary Gas Flow (Ar, L·min^{-1})	46	32	58	58

Table 1. *Cont.*

Parameters	NiCrAlY Bond Coat	YSZ Coat	GYYSZ Coat	Composite Coat
Secondary Gas Flow (H2, L·min^{-1})	6	8	16	12
Feed Rate (g·min^{-1})	84	64	46	59
Spray Distance (mm)	90	90	70	80
Step Size (mm)	4	2	2	2
Spray Gun Traverse Speed (mm·s^{-1})	1000	1000	1000	1000

Phase analysis was conducted using a BRUKER D8 X-ray diffractometer (XRD) with Cu target Kα radiation (λ = 1.5418 Å) at 40 kV and 40 mA. The 2θ scanning range was set from 20° to 90°, with a scanning speed of 5°/min and a step size of 0.02°. The surface and cross-sectional morphology of the coatings, as well as the elemental composition, were observed using a Zeiss GeminiSEM 300 field emission scanning electron microscope (SEM) (Zeiss, Jena, Germany) and a HITACHI S-3400N tungsten filament SEM (Hitachi, Tokyo, Japan), equipped with an energy dispersive X-ray spectrometer (EDS). In the BSE mode of the SEMs, 10 random positions were selected at 500× magnification to observe morphology, and the images were processed using ImageJ software (ImageJ 1.54i) to calculate the porosity of the samples. Ten points were chosen for statistical analysis, with the average value representing the final porosity of the coating. Within ImageJ, five random lines were drawn on SEM grain images at 5000× magnification to measure the true lengths and count the grains intercepted by the lines, allowing for the calculation of average grain size. Thermal conductivity and thermal expansion coefficients of the coating materials were tested using a Netzsch LFA 457 laser flash apparatus and a DIL 402CL dilatometer (Netzsch, Selb, Germany).

Samples were heat-treated using a Nabertherm N61/H atmosphere furnace (Nabertherm, Lilienthal, Germany). The thermal shock testing method for the TBCs involved photographing and weighing TBC-1, TBC-2, and TBC-3 before the experiment. Once the furnace temperature reached 1150 °C, the furnace door was quickly opened to insert the samples, which were held for 5 min. The samples were then rapidly removed and plunged into 25 °C deionized water for quick cooling, soaking for 2 min. After drying, the samples were photographed and weighed again, completing one cycle. This procedure was repeated until a 20% coating delamination was observed, defining the number of cycles at which this occurred as the thermal shock life. Three samples were taken for each set of TBC parameters to undergo thermal shock and water quenching tests. and the average value calculated from these tests represented the average thermal shock life.

3. Results and Discussion

3.1. Microstructural and Property Analysis of GYYSZ Coating

Figure 2 presents the macroscopic photograph, surface morphology, and the EDS elemental distribution map of the sprayed GYYSZ ceramic coating. In Figure 2a, the coating surface appears to be smooth and milky white, matching the color of the ceramic powder. Figure 2b shows no obvious spherical structure of the GYYSZ ceramic powder, indicating that the particles melted in the high-temperature environment of APS. The magnified view reveals two distinct areas on the coating surface: a smooth, fully molten region and a semi-molten region where ceramic powder remains partially unmelted. The height difference between the semi-molten and fully molten regions is due to the varying kinetic energy of the powder particles as they are propelled by the high-speed flame flowing from the nozzle. This variation affects their final state upon impact with the substrate or coating. Ceramic powders with a higher degree of melting exhibit liquid characteristics upon impact, causing them to splash and spread out, forming a layered coating. This type of coating is fully molten and demonstrates good density and adhesion to the substrate. However, during the deposition of these coatings, quenching stresses can occur, leading to the formation of microcracks, as illustrated in Figure 2c. In contrast, ceramic powders with a lower degree of melting tend to bond and stack upon impact with the substrate or may fall off, resulting in increased porosity in the coating. The EDS elemental mapping conducted at the position indicated in Figure 2b reveals that the elements Gd, Yb, Y, and Zr are uniformly distributed throughout the GYYSZ ceramic coating.

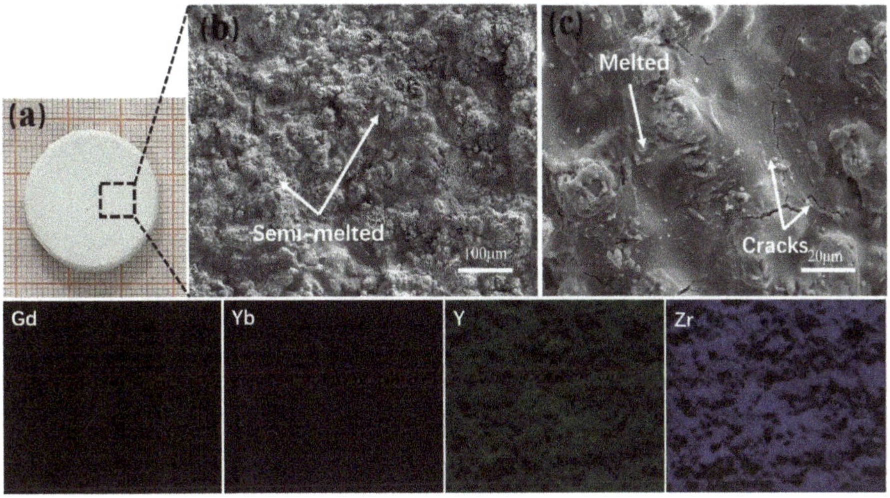

Figure 2. (**a**) The GYYSZ ceramic coating macroscopic photograph; (**b**,**c**) the surface morphology; and the EDS mapping.

Figure 3 presents the XRD patterns of YSZ and GYYSZ in their powdered forms, their as-sprayed states, and following heat treatment at various temperatures. It can be seen from Figure 3a that the YSZ ceramic powder and coating primarily consist of the metastable tetragonal phase t′-ZrO_2, with the powder state containing a significant amount of monoclinic phase m-ZrO_2 [17]. In the sprayed state, the intensity of the m-ZrO_2 characteristic peaks is reduced due to the phase transformation of the ceramic powder at the high temperatures of APS, leading to the formation of t′-ZrO_2. The rapid cooling after spraying restricts the phase transformation from t′ to t to m. Figure 3(a1) displays the XRD patterns of the YSZ coating following heat treatment at various temperatures. After treatment at 1100 °C, the m-ZrO_2 phase in the sprayed YSZ coating gradually disappears. However, after heat treatment, a new phase of Y_2O_3 appears at approximately $2\theta \approx 29°$, indicating that some of the stabilizer Y_2O_3 precipitates from the YSZ above 1100 °C. After heat treatment at 1300 °C, the monoclinic phase m-ZrO_2 reappears at approximately $2\theta \approx 28°$, suggesting that a partial phase transformation of the t′ phase in YSZ has occurred. The phase transformation of t′-ZrO_2 at high temperatures results in the formation of a three-phase mixture, which includes a Y-poor tetragonal phase ZrO_2 [32], a Y-rich cubic phase c-ZrO_2, and a tetragonal phase ZrO_2. Among them, the Y-poor tetragonal phase ZrO_2 has a higher tetragonality and is less able to maintain its metastable structure, gradually transforming into the monoclinic phase as the temperature decreases. The Y-rich c-ZrO_2 is retained, which explains the presence of the diffraction peak for c-ZrO_2 at approximately $2\theta \approx 74°$ in the XRD pattern of YSZ after heat treatment at 1300 °C. To assess the extent of phase transformation in YSZ after heat treatment at 1300 °C, the XRD patterns in the range of 2θ = 72.5°–75° were analyzed, as shown in Figure 3(a2). In this range of diffraction angles, the ZrO_2 lattice exhibits only the tetragonal and cubic phases. The t′-ZrO_2 features characteristic peaks at (004) and (400). In this analysis, curve fitting of the diffraction patterns allows for the determination of peak integral intensities, which can then be used to calculate the molar ratio of cubic ZrO_2 to tetragonal ZrO_2. By combining the integral intensities of the monoclinic and tetragonal phases in the range of 2θ = 27.5°–32°, the contents of each ZrO_2 phase in YSZ after heat treatment at 1300 °C can be calculated using Formulas (1)–(3).

$$\frac{M_m}{M_c+M_t} = 0.82\left(\frac{I_{(\bar{1}11)_m} + I_{(111)_m}}{I_{(111)_{c,t}}}\right) \tag{1}$$

$$\frac{M_c}{M_t} = 0.88\left(\frac{I_{(400)_c}}{I_{(004)_t} + I_{(400)_t}}\right) \tag{2}$$

$$M_m + M_c + M_t = 1 \tag{3}$$

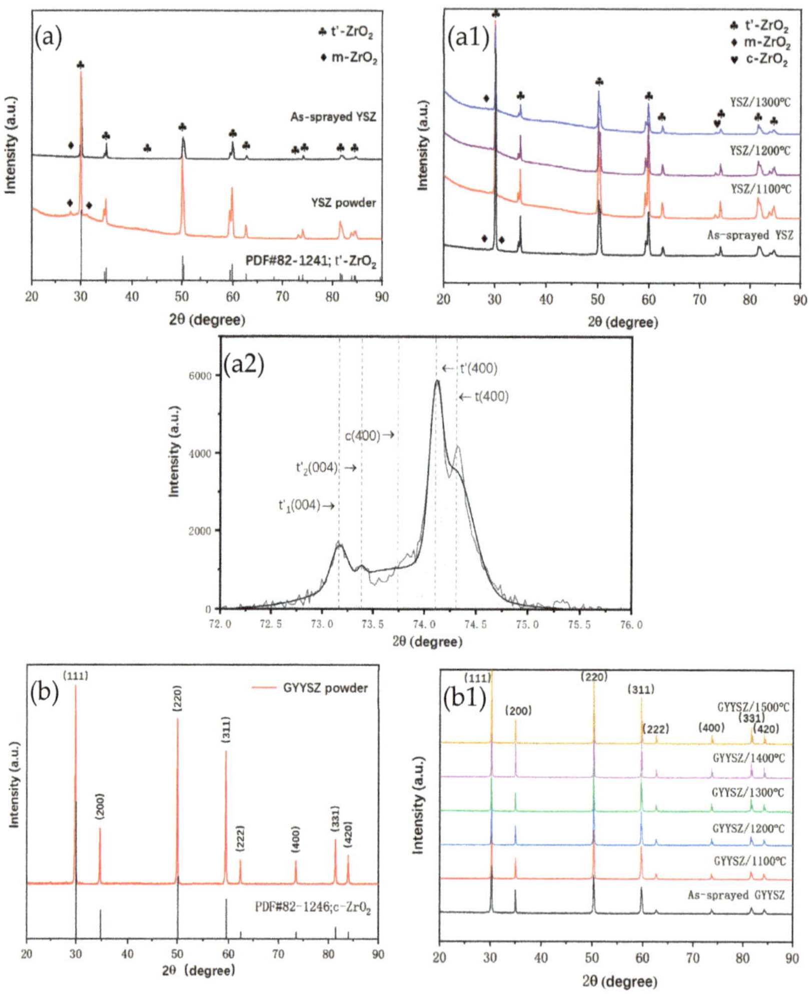

Figure 3. (**a**) The XRD patterns of the YSZ powder and coating; (**a1**) the XRD patterns of the YSZ ceramic coating after heat treatment for 5 h at different temperatures; (**a2**) the XRD patterns and fitting curves in the 2θ = 72.5°–75° range of the YSZ coating after heat treatment at 1300 °C; (**b**) the XRD patterns of GYYSZ in the sprayed state and powder state; (**b1**) the XRD patterns of the GYYSZ ceramic coating after heat treatment for 5 h at different temperatures.

M_m, M_c, and M_t represent the molar fractions of the monoclinic, cubic, and tetragonal phases, respectively, while I denotes the integral intensities of the corresponding characteristic peaks. After calculations, the contents of cubic c-ZrO_2 and tetragonal t-ZrO_2 in YSZ treated at 1300 °C were found to be 9.12% and 18.52%, respectively.

Figure 3b shows the XRD patterns of GYYSZ in powder form, in its sprayed state, and after heat treatment at different temperatures. The diffraction peak angles of the sprayed GYYSZ shifted approximately 0.1° towards higher angles, indicating that the GYYSZ coating has a smaller lattice constant compared to the powder form.

$$2dsin\theta = n\lambda \quad (4)$$

The refractive index n and wavelength λ are constant, and the diffraction angle θ increases as the interplanar spacing d decreases. GYYSZ involves the substitution of Zr sites in the ZrO_2 lattice by Y^{3+}, Gd^{3+}, and Yb^{3+} ions. The ionic radii of Y^{3+}, Gd^{3+}, Yb^{3+}, and Zr^{4+} are shown in Table 2, where $r(Gd^{3+}) > r(Y^{3+}) > r(Yb^{3+}) > r(Zr^{4+})$. Doping ZrO_2 with these three rare earth cations leads to lattice distortion, resulting in an increase in interplanar spacing. At approximately $2\theta \approx 29.2°$, a weak Y_2O_3 diffraction peak appears, indicating the presence of trace Y_2O_3 precipitation. This further confirms that the overall shift to higher angles in the sprayed state is due to the precipitation of rare earth cations during the spraying process. Figure 3(b1) shows the XRD patterns of the coating after heat treatment. After treatment at 1500 °C for 5 h, the phase structure of GYYSZ remains consistent with the sprayed state, which is the c-ZrO_2. This indicates that GYYSZ exhibits better high-temperature structural stability compared to YSZ materials.

Table 2. The ionic radii of the four cations in GYYSZ [33].

Ionic	Y^{3+}	Gd^{3+}	Yb^{3+}	Zr^{4+}
Ionic radii (nm)	0.102	0.105	0.099	0.084

Figure 4 shows the surface morphology images of GYYSZ coatings after heat treatment at different temperatures. As illustrated, the surface of the GYYSZ coating distinctly exhibits a distribution of semi-melted ceramic powder particles and fully melted ceramic layers, which is evident throughout the entire heat treatment process. The semi-melted region is rough and distinctly granular, with a scattering of various-sized pores. In contrast, the fully melted region is smooth and even, featuring a certain number of microcracks. These microcracks are a result of residual stresses induced during the cooling process after spraying. Due to the relatively low fracture toughness of GYYSZ ceramics, microcracks can initiate in the coating under the influence of residual stress. The unmelted nanoparticles in the coating are shown in Figure 4(S1). The nanoparticles exhibit a regular shape, mostly consisting of irregularly sized polyhedra. After heat treatment at different temperatures for 5 h, the shapes of the nanoparticles are depicted in Figure 4(S2–S6), where the edges of the nanoparticles gradually become rounded, and the bonding rate between the layers increases. Comparing the sintering behavior of GYYSZ coatings at various temperatures ranging from 1100 °C to 1500 °C, the sintering process gradually intensifies. Above 1300 °C, some microcracks progressively disappear during sintering.

Figure 5 illustrates the grain distribution of GYYSZ coatings following heat treatment at various temperatures. In the as-sprayed state, no visible grains are present in the GYYSZ coating. Slight sintering occurs, and grains can be observed to be bonding to each other, after heat treatment at 1100 °C. The grain size grows, the grain texture becomes clearer, and the number of microcracks diminishes with increasing heat treatment temperature. Table 3 presents the grain sizes of GYYSZ coatings following heat treatment at various temperatures. At 1100 °C, the coating begins to sinter, and the average grain size is 840 nm. As the temperature increases, the grain size gradually enlarges, and at 1500 °C, significant sintering occurs, with the average grain size reaching 4657 nm, which is 5.5 times the grain size at 1100 °C.

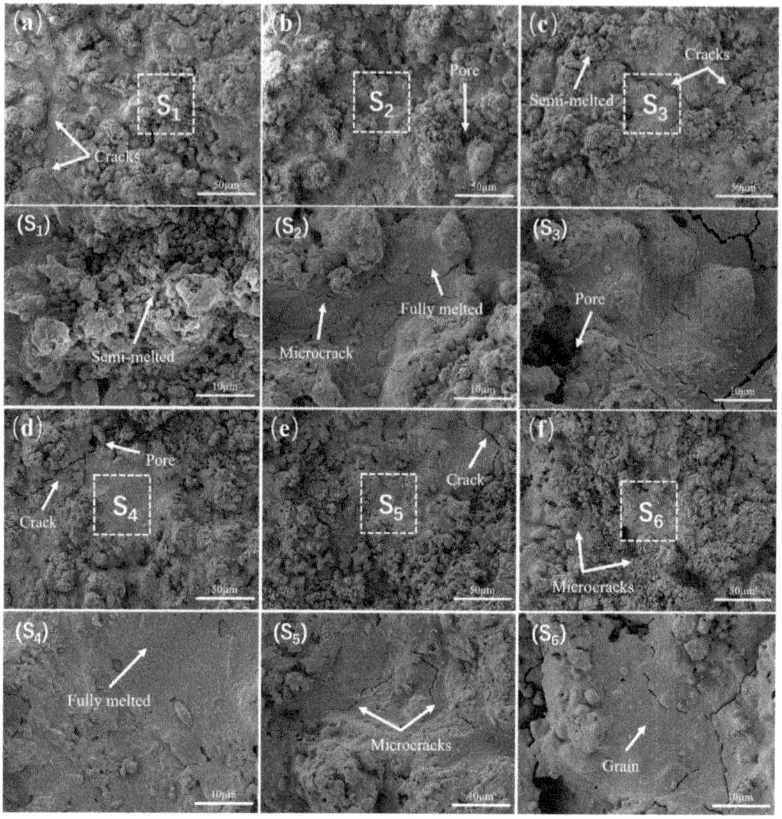

Figure 4. The surface morphology images of GYYSZ coatings after heat treatment for 5 h: (**a**) as-sprayed; (**b**) 1100 °C; (**c**) 1200 °C; (**d**) 1300 °C; (**e**) 1400 °C; and (**f**) 1500 °C.

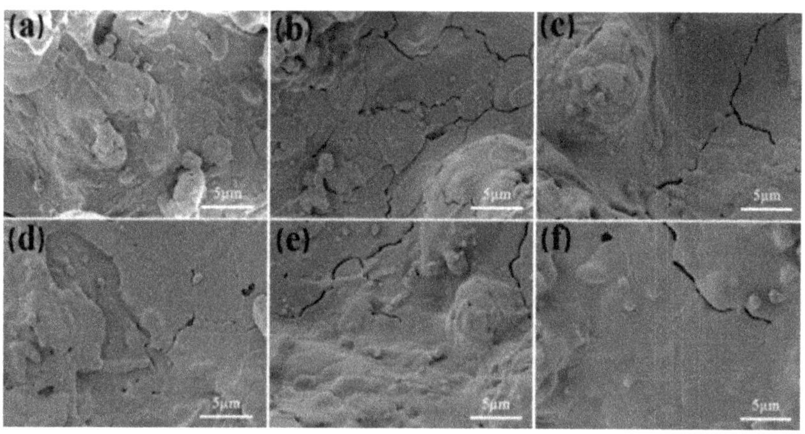

Figure 5. The grain distributions of GYYSZ coatings after heat treatment for 5 h: (**a**) as-sprayed; (**b**) 1100 °C; (**c**) 1200 °C; (**d**) 1300 °C; (**e**) 1400 °C; (**f**) 1500 °C.

Table 3. The changes in grain size of GYYSZ coatings after heat treatment at different temperatures.

Temperature (°C)	1100	1200	1300	1400	1500
Average grain size (nm)	843	991	1353	1721	4657

Figure 6 depicts the cross-sectional images of GYYSZ coatings following heat treatment at various temperatures. As seen in Figure 6a, the high-temperature sintered coatings exhibit significant porosity. In the cross-section, pores combine with microcracks, forming extensive porous regions. The interior of the pores has a rough texture, with some partially unmelted ceramic powder particles present. These pores result from the stacking of incompletely melted powder during the spraying process. The number of microcracks is relatively reduced, and the coating exhibits slight sintering, but the pore diameter has not significantly changed, in Figure 6b. This indicates that, in the initial stage of sintering, the contact area between particles increases. As the heat treatment temperature rises, the pore diameter decreases, the roundness of the pores improves, and the irregular pores gradually develop into holes. At 1300 °C, the porosity of the coating significantly decreases, and the degree of densification increases, indicating that the densification has entered a quasi-steady state. In the final stage of sintering, the large pores in the coating begin to shrink until the holes close. At this stage, the sintering process is primarily driven by sintering stress caused by surface tension, and it lasts for a relatively long time [34]. Table 4 presents the cross-sectional porosity of GYYSZ coatings following heat treatment. As seen in the table, when the treatment temperatures are 1300 °C, 1400 °C, and 1500 °C, the porosities are 10.5%, 11.5%, and 9.9%, respectively. These porosity values are lower than those of the coatings after heat treatment at 1100 °C and 1200 °C under the same annealing time, indicating that higher sintering temperatures accelerate the densification process of the coatings [35]. The porosity no longer decreases at 1300 °C, suggesting that the GYYSZ coating has entered a quasi-steady state of densification. A certain level of porosity is beneficial for reducing phonon-mediated heat transfer, thereby lowering the thermal conductivity of the coating. The fact that the coating maintains a certain porosity even at 1500 °C indicates that GYYSZ has good sintering resistance and maintains structural stability at elevated temperatures.

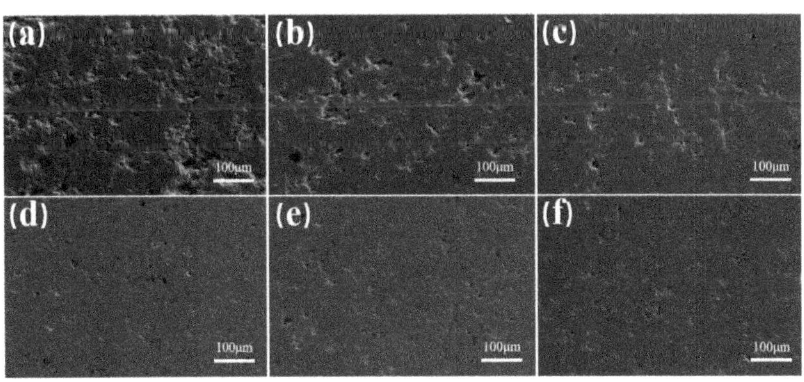

Figure 6. The cross-sectional morphology of GYYSZ coatings after heat treatment at different temperatures: (**a**) as-sprayed; (**b**) 1100 °C; (**c**) 1200 °C; (**d**) 1300 °C; (**e**) 1400 °C; (**f**) 1500 °C.

Table 4. The porosity of GYYSZ coatings in the as-sprayed state and after heat treatment for 5 h.

Temperature (°C)	As-Sprayed	1100	1200	1300	1400	1500
Porosity (%)	21	14.8	13.1	10.5	11.5	9.9

3.2. Design and Performance Study of Gradient Thermal Barrier Coatings

As noted in Section 3.1, GYYSZ exhibits excellent high-temperature anti-sintering capability and high-temperature phase stability, making it a thermal barrier coating material with great development potential. However, the considerable difference in thermal expansion coefficients in GYYSZ and the bonding layer can lead to significant thermal stresses developing within the coating during thermal cycling. When thermal stresses within the coating accumulate to a critical level, they can trigger the initiation and propagation of cracks, jeopardizing the lifespan of the coating. This study designed a composite coating consisting of a 1:1 weighted percentage mixture of YSZ and GYYSZ as an intermediate layer, facilitating thermal expansion transition and improving the thermal cycling durability of the coating.

Figure 7 shows the curves of the thermal conductivity and the thermal expansion coefficient for YSZ, GYYSZ, and the composite material as a function of temperature. All three materials exhibit higher thermal conductivity at low temperatures and lower thermal conductivity at high temperatures, which aligns with the phonon scattering model. At elevated temperatures, increased phonon scattering reduces the average phonon mean free path, resulting in a decrease in the thermal conductivity of the samples. The thermal conductivity of YSZ is higher than that of GYYSZ and the composite material across the entire temperature range from room temperature to 1000 °C. YSZ's thermal conductivity continuously decreases with increasing temperature, with a maximum value of 3.03 $W \cdot m^{-1} K^{-1}$ at room temperature and a minimum of 2.39 $W \cdot m^{-1} K^{-1}$ at 1000 °C. This is approximately 4/5 higher than the thermal conductivity of GYYSZ at the same temperature (1.35 $W \cdot m^{-1} K^{-1}$). The introduction of more dopant ions in GYYSZ creates a greater number of oxygen vacancies, which affects the average free path of phonons [36]. The thermal conductivity of the composite material falls between that of YSZ and GYYSZ. Similarly to GYYSZ, the thermal conductivity of the composite material follows a trend of initially increasing, then decreasing, and rising again as the temperature increases. At 200 °C, the thermal conductivity of the composite material reaches its maximum value of 1.99 $W \cdot m^{-1} K^{-1}$, and at 700 °C, it reaches its minimum value of 1.79 $W \cdot m^{-1} K^{-1}$. After that, the thermal conductivity gradually increases. This increase in thermal conductivity at higher temperatures is due to the densification of the sample at that temperature. At the same temperature, the minimum thermal conductivity of the composite material is still about 1/5 higher than that of GYYSZ. Overall, the thermal conductivity of the composite material remains between 1.79 $W \cdot m^{-1} K^{-1}$ and 1.99 $W \cdot m^{-1} K^{-1}$ from room temperature to 1000 °C, lying between the thermal conductivities of YSZ and GYYSZ. Figure 7b shows the curve of the thermal expansion coefficient for YSZ, GYYSZ, and the composite material as a function of temperature. In the temperature range of 100–1200 °C, the thermal expansion coefficient of the composite material increases with rising temperature. Below 300 °C, its value rapidly increases from $7.05 \times 10^{-6} K^{-1}$ to $10.27 \times 10^{-6} K^{-1}$. In contrast, the thermal expansion coefficient of YSZ increases only slightly in the 100–300 °C range, from $9.33 \times 10^{-6} K^{-1}$ to $10.88 \times 10^{-6} K^{-1}$. This indicates that YSZ has a slower rate of increase in its thermal expansion coefficient, leading to a more gradual growth in thermal strain. This more stable behavior during cooling helps to effectively reduce the generation of thermal stress, thereby lowering the likelihood of coating delamination and failure during thermal shock [37]. At temperatures ranging from 100 to 800 °C, the thermal expansion coefficients follow the order of YSZ > composite material > GYYSZ. At 800 °C, the thermal expansion coefficients of the composite and YSZ become similar, approximately $10.7 \times 10^{-6} K^{-1}$. As the temperature increases, the thermal expansion coefficient of the composite increases in sync with GYYSZ. Therefore, the composite material can be used as an intermediate layer to bridge GYYSZ and YSZ, facilitating a transition in thermal expansion coefficients among the components of the thermal barrier coating and addressing the issue of thermal mismatch caused by the low thermal expansion coefficient of GYYSZ.

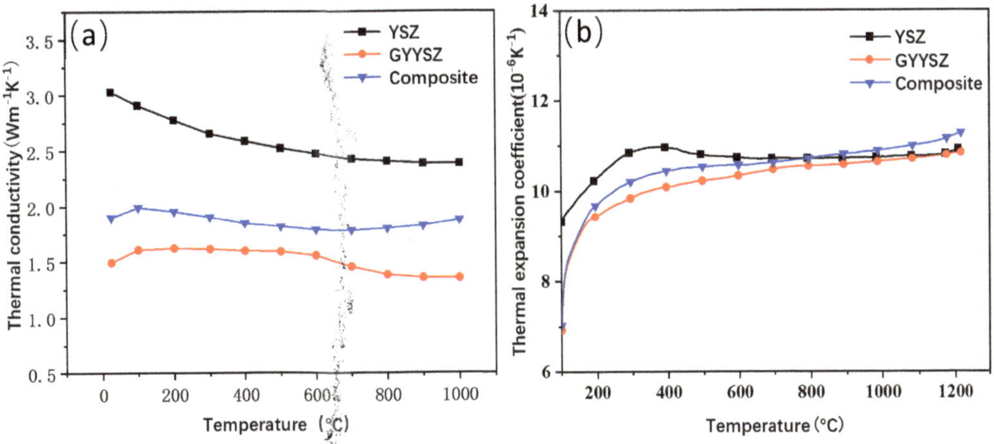

Figure 7. (**a**) The thermal conductivity; (**b**) the thermal expansion coefficient for YSZ [38], GYYSZ, and the composite material.

Figure 8 shows the cross-sectional images of three different TBCs prepared using APS technology. GYYSZ is deposited on the top layer of the TBC. TBC-1 serves as the control group without a transition layer. In TBC-2 and TBC-3, YSZ acts as the transition layer. This is because YSZ has a relatively high thermal expansion coefficient and is deposited between the GYYSZ and the bonding layer (NiCrAlY) to establish a thermal expansion coefficient transition. In TBC-3, a composite coating is sprayed between the YSZ and GYYSZ layers, further enhancing the transition in thermal expansion coefficients. Figure 8a shows the cross-sectional microstructure and localized elemental distribution of TBC-1. The thickness of the GYYSZ layer is approximately 360 μm. The GYYSZ layer contains a significant number of pores and cracks, with the cracks predominantly oriented vertically. The porosity is about 20.97%. The metallic bonding layer and NiCrAlY are well-melted, and the interface between the ceramic layer and the bonding layer is distinct, with the boundary appearing as an irregular curve. This irregularity helps increase the contact area between the ceramic and bonding layers, thereby enhancing the bonding strength. Nevertheless, because of the significant porosity of the GYYSZ layer, pores are also present at the interface edges, which may affect the bonding strength. From the EDS elemental mapping results, it can be observed that the distribution of Gd, Yb, Y, and Zr elements is uniform. The GYYSZ top layer has a thickness of approximately 120 μm and the YSZ intermediate layer is around 240 μm thick in Figure 8b. The porosity of the as-sprayed YSZ layer is about 7.59%, and transverse microcracks are primarily distributed within the coating. The interface between the bonding layer and the ceramic layer is clearly defined, indicating good adhesion. However, it is difficult to distinguish the boundary between GYYSZ and YSZ in the SEM images due to their similar compositions and contrast. By performing an EDS elemental mapping on the selected region S2, the boundary between the coatings can be observed, showing that GYYSZ contains higher concentrations of Gd and Yb, while YSZ has a higher Zr content. Figure 8c shows the SEM image of the graded thermal barrier coating and the corresponding local EDS element distribution. It can be observed that the thickness of the three ceramic layers is approximately 120 μm. There are a certain number of pores in the cross-section of both the GYYSZ and composite layers, and the GYYSZ top layer exhibits some vertical cracks. EDS mapping of the S3 region reveals an alternating distribution of GYYSZ and YSZ within the composite layer. Additionally, the interfaces between the ceramic layers are clearly visible, with the boundary lines being irregular and winding, showing no significant cracks, indicating good bonding between the layers of the graded thermal barrier coating.

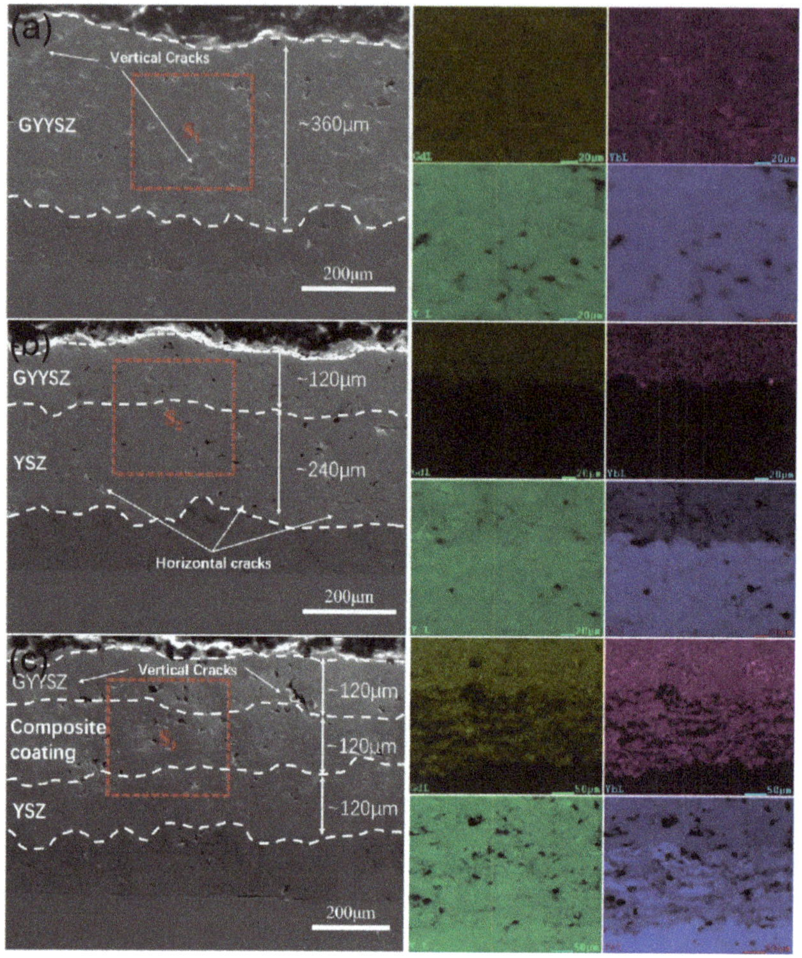

Figure 8. The cross-sectional SEM images and EDS element distribution maps for (**a**) TBC-1; (**b**) TBC-2; and (**c**) TBC-3.

Figure 9 illustrates the average thermal shock times of the three types of TBCs. The average thermal shock time is the shortest for TBC-1, at just 42.3 cycles. In contrast, TBC-2 shows a significant increase in thermal shock time, reaching an average of 136.3 cycles, which is 3.2 times that of TBC-1. This indicates that the dual-ceramic layer structure offers significantly better thermal shock resistance. Due to the low thermal expansion coefficient of GYYSZ, thermal mismatch occurs between the ceramic layer and the bonding layer in TBC-1. Depositing a layer of YSZ between GYYSZ and the bonding layer, the higher thermal expansion coefficient of YSZ at mid to low temperatures generates less thermal stress during thermal cycling compared to TBC-1, thereby extending the thermal cycling times of the coating. The average thermal shock time of TBC-3 is 246.3 cycles, nearly six times that of TBC-1 and 80% more than TBC-2. This indicates that adding a transition layer can significantly enhance the thermal shock lifespan of the TBC. The thermal shock test results show that modifying the compositional gradient of the TBC structure can enhance its thermal shock lifespan. Furthermore, adding a transition layer in coatings prone to thermal mismatch can create a gradient in thermal expansion coefficients, significantly improving thermal shock resistance.

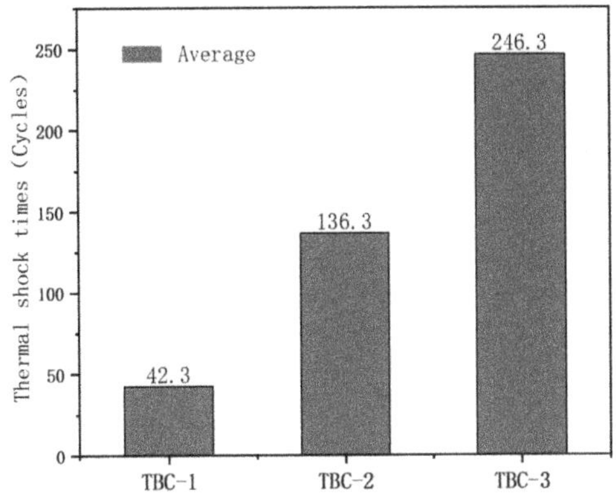

Figure 9. Average thermal shock times of three different structural TBCs.

Figure 10 shows photographs of the three types of TBCs during thermal cycling experiments at different times of cycles, focusing on four stages: at the as-sprayed stage, at initial spalling, during ongoing spalling, and at thermal shock failure. The TBC failure mode is mainly in the form of ceramic layer peeling. TBC-1 exhibits large-scale spalling of the ceramic layer; TBC-2 shows failure in the center area with a failure area reaching 20%; TBC-3 demonstrates spalling failures at both the edges and the center of the ceramic layer. As shown in Figure 10(a1–a4), TBC-1 shows small spalling at the edges of the ceramic layer after 13 thermal shock cycles. As the number of cycles increases, the spalled area gradually enlarges, and by the 38th cycle, large areas of the ceramic layer have spalled off, displaying a gray color, leading to the termination of the experiment. When TBC-1 has completely failed, neither TBC-2 nor TBC-3 show any signs of ceramic layer spalling. However, after 46 thermal cycles, as shown in Figure 10(b2), visible spalling areas appear on the surface of TBC-2, with the rate of ceramic spalling increasing rapidly. Following an additional 9 water quenching cycles, the spalled area reaches 7.6% (Figure 10(b3)). Subsequently, the spalling rate decreases, and the spalled area slowly increases with the number of thermal shock cycles, ultimately failing after the 153rd cycle. Figure 10(c2–c6) show the spalling process of TBC-3. After 116 thermal quenching cycles, small spalling areas appear at the center of the coating, with the spalled area expanding slowly as the number of thermal cycles increases. After 156 thermal shock cycles, spalling is observed at the edges, but by the 179th thermal cycle, the spalled area still accounts for less than 1%. TBC-3 fails after 313 thermal shock cycles, and the failure modes of the two failure regions are similar to those of TBC-2.

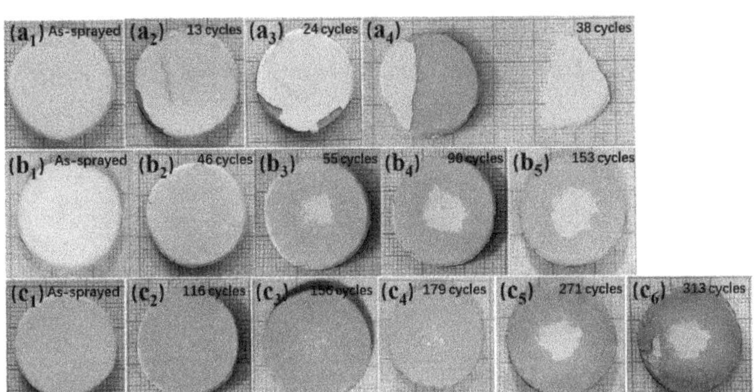

Figure 10. Macro photographs of three TBCs after varying times of thermal cycles: (**a1–a4**) TBC-1; (**b1–b5**) TBC-2; (**c1–c6**) TBC-3.

Figure 11 shows images of TBC-1 failure. As seen in Figure 11(a1), the surface of the fallen layer exhibits cracks. This occurs during heating, where the high thermal expansion coefficient of the bonding layer exerts compressive stress on the ceramic layer, while during cooling, the ceramic layer experiences tensile stress. In the thermal shock cycling experiment, the coating experiences a repetitive cycle of tensile and compressive stresses, which induces the initiation, and propagation of microcracks. After the ceramic layer of TBC-1 delaminates, the exposed area appears gray, indicating that the delamination occurs near the interface between GYYSZ and the bonding layer. The EDS elemental mapping results for the failed regions are shown in Figures 11(b1) and 12(b2). The surface of the failed area is rough, and the EDS analysis indicates that, in addition to Gd, Yb, Zr, and Y, there is a significant presence of Ni, Cr, and Al elements in the areas where Zr is missing. Ni, Cr, and Al are the main components of the bonding layer. This suggests that the failure of the coating occurred at the interface between the ceramic layer and the bonding layer. Figure 11c shows the cross-sectional view of TBC-1 after failure. The coating surface has lost most of its ceramic layer due to thermal shock, as seen in Figure 11(c1). Although no through-thickness transverse cracks are observable in the area where the ceramic layer has detached, small transverse cracks are still visible. This is due to the significant thermal mismatch stress generated during thermal shock, caused by the large difference in thermal expansion coefficients between the ceramic layer and the bonding layer. This leads to the initiation of transverse cracks in the low-toughness ceramic, which propagate during thermal cycling [39,40]. TBC-1 has undergone few thermal shock cycles, so the accumulated stress in the ceramic layer has not yet reached the bonding strength between the ceramic materials. The competition between longitudinal cracks and lateral interface cracks is dominated by the development of lateral interface cracks, ultimately leading to the delamination observed in Figure 11(c1). The final failure of TBC-1 is attributed to the through-thickness transverse cracks caused by thermal mismatch stress.

Figure 12 shows the failure images of TBC-2. Figure 12(a1) displays the BSE photograph of the unspalled area of TBC-2. In the image, a network of surface cracks can be observed on the coating surface. When the thermal mismatch coefficients of the components in the TBC are significant, the ceramic layer experiences thermal mismatch stress. During cooling, the surface of the ceramic layer is subjected to substantial tensile stress, inducing cracks. These cracks initiate at the center of the sample and propagate downward vertically during the cyclical thermal cycling process. From Figure 12b, There is an obvious high and low difference on the coating [41]. The layer is peeling step by step, and obvious interlayer cracks can be observed, which is the result of transverse cracks and longitudinal cracks between the ceramic layer and peeling successively. An EDS mapping of the lowest ceramic layer indicated that this area is still GYYSZ. This suggests that the initial peeling occurred within the GYYSZ layer, followed by the formation of inter-layer cracks that converged with longitudinal cracks, resulting in localized peeling. Subsequent peeling events continued to occur within this layer. Figure 12c shows the cross-sectional structure of the unspalled area and the spalled area of TBC-2. In the unspalled area, both transverse and longitudinal cracks are present in the ceramic layer. The transverse cracks are mainly located at two interfaces: one is between GYYSZ and YSZ, and the other is between YSZ and NiCrAlY. The former has developed into through-type cracks, posing a risk of spalling at any moment, while the latter is still in the stage of expansion. There are two reasons for this situation: First, GYYSZ and YSZ tend to generate larger thermal mismatch stresses, leading to faster crack propagation at the GYYSZ/YSZ interface under the same number of thermal cycles. Second, due to the inherent properties of YSZ, when cracks are generated and propagate at the coating interface under tensile stress, the stress at the crack tip can induce a martensitic phase transformation from the t′ to the m phase in YSZ, primarily from the t′-ZrO_2 phase. This phase transformation induces a volume expansion that partially offsets the tensile stress as compressive stress, enhancing toughness and effectively delaying the propagation of cracks at the YSZ/NiCrAlY interface. The addition of the YSZ layer replaces the direct contact between GYYSZ and the bonding layer, effectively alleviating interface stress, delaying coating spallation, and enhancing the lifespan of the coating. During the cooling process, the tensile stress on the surface of the GYYSZ coating increases with the number of quenching cycles, leading to the initiation and extension of longitudinal cracks at the interface. The stress is more concentrated at the center of TBC-2, causing cracks to preferentially develop there, resulting in the initial spallation of the coating that gradually extends towards the edges of the sample [42]. Additionally, the further away from the center of the sample, the lower the quenching stress, which in turn reduces the spallation rate. This explains why the area of spallation in TBC-2 initially grows rapidly after the onset of spallation, but then slows down. The cracks within the YSZ layer develop slowly, ultimately forming the transverse and longitudinal cracks shown in Figure 12c, which trigger the spallation of the YSZ ceramic. Thus, the high thermal expansion coefficient of YSZ helps to mitigate the thermal mismatch stress between the GYYSZ and the bonding layer. The interplay of interlayer cracks, interface cracks, and surface cracks in the GYYSZ/YSZ structure collectively leads to the failure of the coating.

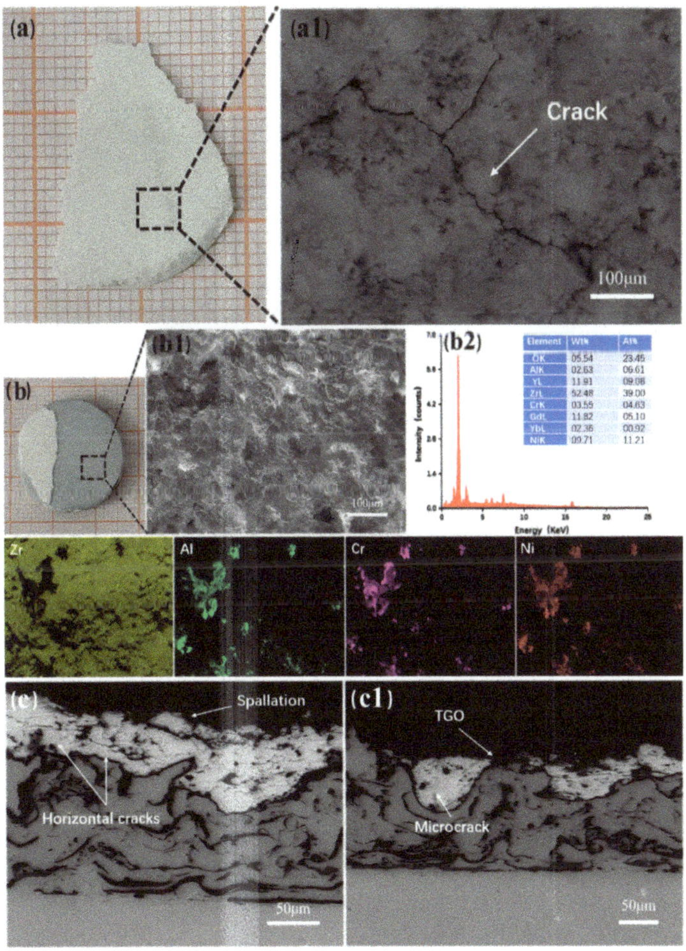

Figure 11. (**a**) The failure surface morphology of TBC-1; (**a1**) BSE diffraction image; (**b**) failure photographs of TBC-1; (**b1**) surface morphology of the failure location; (**b2**) EDS results and elemental distribution; and (**c**,**c1**) the cross-sectional microstructure of two locations of TBC-1 after thermal shock failure.

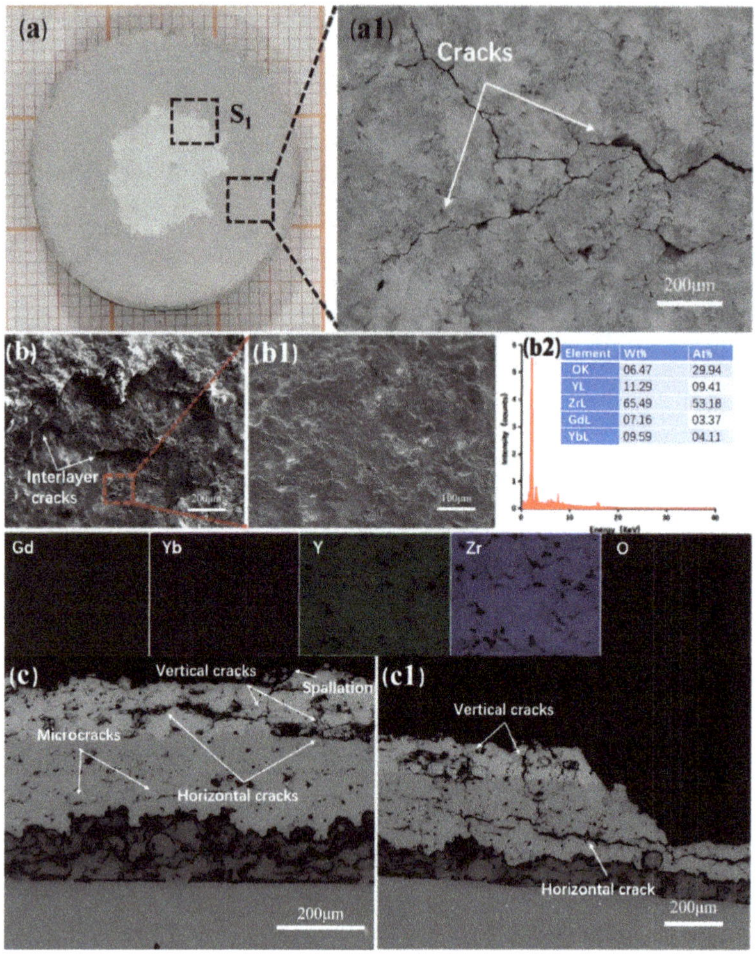

Figure 12. (**a**) Failure surface morphology of TBC-2; (**a1**) BSE image; (**b**) failure image of TBC-1; (**b1**) surface morphology of failure area; (**b2**) EDS mapping; (**c**,**c1**) cross-sectional microstructure of TBC-1 at two locations after thermal shock failure.

Figure 13 shows the failure images of TBC-3. As shown in Figure 13(a1), the unpeeled area exhibits a network of surface cracks, resulting from the release of tensile stress in the thickness direction of the coating. Compared to the surface cracks in TBC-2, the cracks in TBC-3 are finer, which helps delay coating spalling. Compared to the surface cracks in TBC-2, the cracks in TBC-3 are finer, which helps delay coating spalling. Analyzing the different failure characteristics at locations S2 and S3, as shown in Figure 13b,c, reveals the surface morphology and EDS spectra for selected areas. The results indicate that the failure mode at S2 is similar to that of TBC-2, with the peeling edge still consisting of GYYSZ. In contrast, S3 shows significant height differences, indicating direct debonding at this location, and the presence of Ni, Cr, and Al elements in the spectrum confirms that a failure mode similar to that of TBC-1 occurred here.

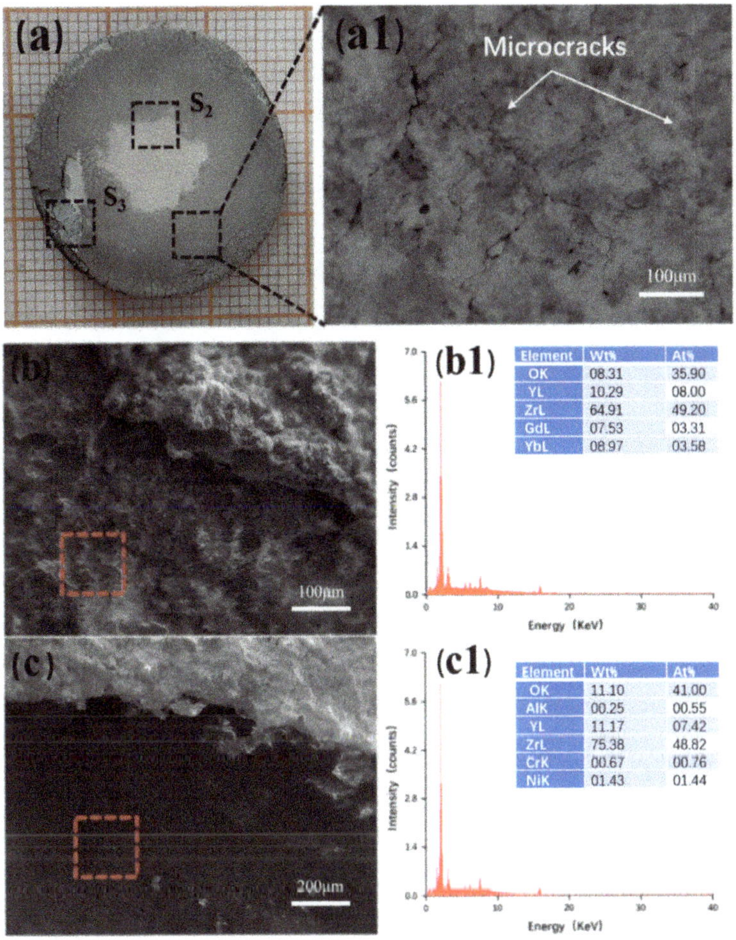

Figure 13. (a) The surface morphology of the failed TBC-3 coating; (**a1**) a BSE image of the non-spalled area; (**b**,**b1**) the surface morphology and EDS spectra at positions S2, and at (**c**,**c1**) S3.

Figure 14 shows cross-sectional images of spalling at the center and edges. In Figure 14a, the thickness of the GYYSZ layer is reduced, and block-like ceramics are observed on the layer, indicating that the spalling occurs within the GYYSZ top layer. There are no obvious longitudinal cracks observed within the composite layer and the YSZ layer, and only minor transverse cracks are present at the interface between NiCrAlY and YSZ. This indicates that optimizing the structure of the thermal barrier coating can significantly reduce the thermal mismatch stress between the components. In the competition between interface cracks and surface cracks, the expansion of surface cracks dominates, leading to the failure and detachment of the coating. In the cross-sectional morphology of the GYYSZ layer, as shown in Figure 14b, numerous microcracks can be observed within the coating. The presence of shorter and higher-density microcracks can reduce the driving force for interface cracks. These microcracks can relieve the thermal mismatch stress at the interface, effectively alleviating interface stress, suppressing the formation and growth of transverse cracks, and thereby delaying coating spallation and enhancing thermal cycling life [43–45]. The coating primarily fails through the slow propagation, coalescence, and spallation of cracks. However, as the number of thermal shock cycles increases, interface cracks gradually extend to the state shown in Figure 14b. Figure 14c,d illustrate the cross-sectional microstructure of the spalled TBC-3. During quenching, the thermal gradient stress at the circular edge of the sample is lower. The high-density surface cracks help to relieve the stress within the ceramic layer, allowing the relatively large difference in thermal expansion coefficients at the YSZ/NiCrAlY interface to become the primary driving force for interface crack propagation, resulting in the formation of penetrating transverse cracks at the interface, which represent the main failure mode.

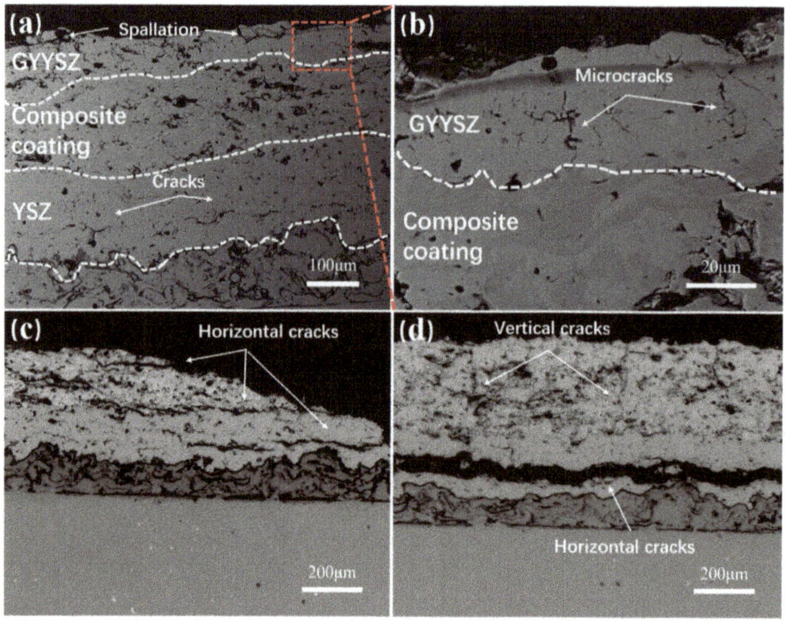

Figure 14. (**a**) The spalled area of TBC-3 after thermal cycling failure; (**b**) a local magnified view; (**c**) the cross-sections of the spall center; the (**d**) edge spall area.

4. Conclusions

(1) GYYSZ powder exhibits a single cubic phase of c-ZrO_2. After undergoing annealing treatments, the GYYSZ coating remains in the cubic phase of c-ZrO_2. With the rise in annealing temperature, the grain size of the GYYSZ coating also increases, reaching a maximum of 4.66 μm. Additionally, the coating retains a certain porosity during annealing, maintaining approximately 9.9% porosity even at 1500 °C, demonstrating a degree of anti-sintering performance.

(2) By comparing the thermal conductivity and thermal expansion coefficients of the three materials, it is found that GYYSZ has a lower thermal conductivity than YSZ and the composite material, reaching its lowest value of 1.35 $W \cdot m^{-1} K^{-1}$ at 1000 °C, which is 44% lower than that of YSZ. Additionally, GYYSZ has the smallest thermal expansion coefficient, while the thermal expansion coefficient of the composite material falls between that of GYYSZ and YSZ in the temperature range of 100–800 °C. This allows the composite material to serve as a transition layer, alleviating the issue of thermal expansion coefficient mismatch. An analysis of the cross-sections of the three TBCs shows that the spraying thickness is consistent with the design, and EDS results reveal a differential distribution of elements, with irregular boundaries and good bonding performance.

(3) TBC-1 failed after an average of 42.3 thermal shock cycles, resulting in large areas of ceramic layer delamination and a significant drop in quality. The failure was caused by a large thermal mismatch coefficient between the ceramic layer and the bonding layer, leading to concentrated thermal mismatch stress that induced the initiation and rapid propagation

of interfacial transverse cracks, ultimately resulting in the formation of through-thickness transverse cracks and the large-scale delamination of the ceramic layer. TBC-2 exhibited a thermal cycling life of 126.3 cycles. The functionally graded coating structure of TBC-3 further optimized the coating design by adding a transition layer, demonstrating excellent thermal shock resistance with an average thermal cycling life of 246.3 cycles, 1.8 times that of TBC-2. The gradient structure effectively reduced thermal mismatch stress, and the fine, dense surface microcracks contributed to a certain degree of toughening. The failure mode of the coating shifted to a mixed mode of block- and point-like delamination.

Author Contributions: Conceptualization, R.L. and C.J.; methodology, C.J. and Y.Z.; formal analysis, W.L.; investigation, F.H.; resources, R.L.; writing—original draft preparation, C.J.; writing—review and editing, C.J. and R.L.; project administration, Z.C.; funding acquisition, R.L. All authors have read and agreed to the published version of the manuscript.

Funding: This research was funded by the National Natural Science Foundation of China Joint Fund Priority Program (U23A20607), the Shanghai Engineering Research Center of Hot Manufacturing (18DZ2253400), and the Shanghai Municipal Industrial Foundation Strengthening Special Project (GYQJ-2023-1-06).

Data Availability Statement: Data are contained within the article.

Conflicts of Interest: The authors declare no conflicts of interest.

References

1. Xie, F.; Li, D.; Zhang, W. Long-Term Failure Mechanisms of Thermal Barrier Coatings in Heavy-Duty Gas Turbines. *Coatings* **2020**, *10*, 1022. [CrossRef]
2. Wu, Y.; Song, W.; Dingwell, D.B.; Guo, H. Silicate ash-resistant novel thermal barrier coatings in gas turbines. *Corros. Sci.* **2022**, *194*, 109929. [CrossRef]
3. Miller, R.A. Current status of thermal barrier coatings—An overview. *Surf. Coat. Technol.* **1987**, *30*, 1–11. [CrossRef]
4. Miller, R.A. Thermal barrier coatings for aircraft engines history and directions. *J. Therm. Spray Technol.* **1997**, *6*, 35–42. [CrossRef]
5. Kumar, V.; Balasubramanian, K. Progress update on failure mechanisms of advanced thermal barrier coatings: A review. *Prog. Org. Coat.* **2016**, *90*, 54–82. [CrossRef]
6. Mauer, G.; Jarligo, M.O.; Mack, D.E.; Vaßen, R. Plasma-Sprayed Thermal Barrier Coatings: New Materials, Processing Issues, and Solutions. *J. Therm. Spray Technol.* **2013**, *22*, 646–658. [CrossRef]
7. Lakiza, S.M.; Grechanyuk, M.I.; Ruban, O.K.; Redko, V.P.; Glabay, M.S.; Myloserdov, O.B.; Dudnik, O.V.; Prokhorenko, S.V. Thermal Barrier Coatings: Current Status, Search, and Analysis. *Powder Metall. Met. Ceram.* **2018**, *57*, 82–113. [CrossRef]
8. Mulla, J.; Kaka, F.; Satapathy, R.K. The emergence of reflective thermal barrier coatings. *Crit. Rev. Solid State Mater. Sci.* **2023**, *49*, 582–606. [CrossRef]
9. Boissonnet, G.; Chalk, C.; Nicholls, J.R.; Bonnet, G.; Pedraza, F. Phase stability and thermal insulation of YSZ and erbia-yttria co-doped zirconia EB-PVD thermal barrier coating systems. *Surf. Coat. Technol.* **2020**, *389*, 125566. [CrossRef]
10. Yang, M.; Zhu, Y.; Wang, X.; Wang, Q.; Ai, L.; Zhao, L.; Chu, Y.; Guo, S.; Hu, J.; Zhang, Q. Preparation and thermophysical properties of Ti^{4+} doped zirconia matrix thermal barrier coatings. *J. Alloys Compd.* **2019**, *777*, 646–654. [CrossRef]
11. Bobzin, K.; Zhao, L.; Öte, M.; Königstein, T. A highly porous thermal barrier coating based on Gd_2O_3–Yb_2O_3 co-doped YSZ. *Surf. Coat. Technol.* **2019**, *366*, 349–354. [CrossRef]
12. Khan, M.; Zeng, Y.; Lan, Z.; Wang, Y. Reduced thermal conductivity of solid solution of 20% CeO_2 + ZrO_2 and 8% Y_2O_3 + ZrO_2 prepared by atmospheric plasma spray technique. *Ceram. Int.* **2019**, *45*, 839–842. [CrossRef]
13. Bahamirian, M.; Hadavi, S.M.M.; Farvizi, M.; Rahimipour, M.R.; Keyvani, A. Phase stability of ZrO_2 9.5Y_2O_3 5.6Yb_2O_3 5.2Gd_2O_3 compound at 1100 °C and 1300 °C for advanced TBC applications. *Ceram. Int.* **2019**, *45*, 7344–7350. [CrossRef]
14. Bobzin, K.; Zhao, L.; Öte, M.; Linke, T.F. Deposition and characterization of thermal barrier coatings of ZrO_2–4 mol.% Y_2O_3–1 mol.% Gd_2O_3–1 mol.% Yb_2O_3. *Surf. Coat. Technol.* **2015**, *268*, 205–208. [CrossRef]
15. Chen, D.; Wang, Q.; Liu, Y.; Ning, X. Investigation of ternary rare earth oxide-doped YSZ and its high temperature stability. *J. Alloys Compd.* **2019**, *806*, 580–586. [CrossRef]
16. Fan, W.; Wang, Z.Z.; Bai, Y.; Che, J.W.; Wang, R.J.; Ma, F.; Tao, W.Z.; Liang, G.Y. Improved properties of scandia and yttria co-doped zirconia as a potential thermal barrier material for high temperature applications. *J. Eur. Ceram. Soc.* **2018**, *38*, 4502–4511. [CrossRef]
17. Chen, D.; Wang, Q.; Liu, Y.; Ning, X. Microstructure, thermal characteristics, and thermal cycling behavior of the ternary rare earth oxides (La_2O_3, Gd_2O_3, and Yb_2O_3) co-doped YSZ coatings. *Surf. Coat. Technol.* **2020**, *403*, 126387. [CrossRef]
18. Guo, L.; Guo, H.; Gong, S.; Xu, H. Improvement on the phase stability, mechanical properties and thermal insulation of Y_2O_3-stabilized ZrO_2 by Gd_2O_3 and Yb_2O_3 co-doping. *Ceram. Int.* **2013**, *39*, 9009–9015. [CrossRef]

19. Bahamirian, M.; Hadavi, S.M.M.; Farvizi, M.; Keyvani, A.; Rahimipour, M.R. Hot corrosion behavior of ZrO_2 $9.5Y_2O_3$ $5.6Yb_2O_3$ $5.2Gd_2O_3$ TBCs in CMAS: CaO-MgO-Al_2O_3-SiO_2. *J. Aust. Ceram. Soc.* **2020**, *57*, 215–224. [CrossRef]
20. Li, Y.; She, Y.; Liao, K. Hot-Corrosion Behavior of Gd_2O_3–Yb_2O_3 Co-Doped YSZ Thermal Barrier Coatings in the Presence of V_2O_5 Molten Salt. *Coatings* **2023**, *13*, 886. [CrossRef]
21. Shen, H.; Lei, Y.; Lv, X.; Luo, Y.; Li, J.; Sun, L.; Wang, L.; Zhang, J.; Wang, J. Effect of Gd doping on phase evolution, mechanical and thermal characteristics of 1La-xGd-2Yb-3.5YSZ solid solutions. *Ceram. Int.* **2024**, *50*, 2987–2994. [CrossRef]
22. Liu, Y.; Zhen, Z.; Wang, X.; Mu, R.; He, L.; Xu, Z. Thermo-physical properties, morphology and thermal shock behavior of EB-PVD thermal barrier coating with DLC YbGdZrO/YSZ system. *Mater. Today Commun.* **2023**, *35*, 106265. [CrossRef]
23. Wang, L.; Wang, Y.; Sun, X.G.; He, J.Q.; Pan, Z.Y.; Wang, C.H. Thermal shock behavior of 8YSZ and double-ceramic-layer $La_2Zr_2O_7$/8YSZ thermal barrier coatings fabricated by atmospheric plasma spraying. *Ceram. Int.* **2012**, *38*, 3595–3606. [CrossRef]
24. Xu, Z.; He, L.; Mu, R.; Zhong, X.; Zhang, Y.; Zhang, J.; Cao, X. Double-ceramic-layer thermal barrier coatings of $La_2Zr_2O_7$/YSZ deposited by electron beam-physical vapor deposition. *J. Alloys Compd.* **2009**, *473*, 509–515. [CrossRef]
25. Guo, H.; Xu, H.; Gong, S. A study on gradient thermal barrier coatings by EB-PVD in a cyclic high-temperature hot-corrosion environment. *J. Mater. Sci.* **2002**, *37*, 5333–5337. [CrossRef]
26. Chen, X.; Gu, L.; Zou, B.; Wang, Y.; Cao, X. New functionally graded thermal barrier coating system based on $LaMgAl_{11}O_{19}$/YSZ prepared by air plasma spraying. *Surf. Coat. Technol.* **2012**, *206*, 2265–2274. [CrossRef]
27. Portinha, A.; Teixeira, V.; Carneiro, J.; Martins, J.; Costa, M.F.; Vassen, R.; Stoever, D. Characterization of thermal barrier coatings with a gradient in porosity. *Surf. Coat. Technol.* **2005**, *195*, 245–251. [CrossRef]
28. Chen, H.; Liu, Y.; Gao, Y.; Tao, S.; Luo, H. Design, Preparation, and Characterization of Graded YSZ/$La_2Zr_2O_7$ Thermal Barrier Coatings. *J. Am. Ceram. Soc.* **2010**, *93*, 1732–1740. [CrossRef]
29. Chen, X.; Zhao, Y.; Fan, X.; Liu, Y.; Zou, B.; Wang, Y.; Ma, H.; Cao, X. Thermal cycling failure of new $LaMgAl_{11}O_{19}$/YSZ double ceramic top coat thermal barrier coating systems. *Surf. Coat. Technol.* **2011**, *205*, 3293–3300. [CrossRef]
30. Li, R.; Xing, Y.; Li, Q.; Cheng, Z.; Guo, L. High-Temperature Oxidation Resistance and Molten Salt Corrosion Study of YSZ, CeYSZ, and YSZ/CeYSZ Thermal Barrier Coatings by Atmospheric Plasma Spraying. *Coatings* **2024**, *14*, 102. [CrossRef]
31. Wu, S.; Zhao, Y.; Li, W.; Liu, W.; Wu, Y.; Liu, F. Optimization of process parameters for thermal shock resistance and thermal barrier effects of 8YSZ thermal barrier coatings through orthogonal test. *AIP Adv.* **2021**, *11*, 065215. [CrossRef]
32. Tsipas, S.A. Effect of dopants on the phase stability of zirconia-based plasma sprayed thermal barrier coatings. *J. Eur. Ceram. Soc.* **2010**, *30*, 61–72. [CrossRef]
33. Shannon, R.D. Revised effective ionic radii and systematic studies of interatomic distances in halides and chalcogenides. *Acta Cryst.* **1976**, *32*, 751–767. [CrossRef]
34. Wang, Y.; Long, Y.; Yang, C.; Liu, Z.; Guo, F. Effect of sintering on stress distribution of thermal barrier coatings for high temperature blade. *Mater. Chem. Phys.* **2024**, *319*, 129402. [CrossRef]
35. Xue, Z.; Zhou, L.; Shi, M.; Zhang, Z.; Byon, E.; Zhang, S. Preparation and sintering behavior of GdYb-YSZ nanostructured thermal barrier coating. *J. Mater. Res. Technol.* **2023**, *26*, 7237–7247. [CrossRef]
36. Zhao, M.; Pan, W.; Wan, C.; Qu, Z.; Li, Z.; Yang, J. Defect engineering in development of low thermal conductivity materials: A review. *J. Eur. Ceram. Soc.* **2017**, *37*, 1–13. [CrossRef]
37. Bogdan, M.; Peter, I. A Comprehensive Understanding of Thermal Barrier Coatings (TBCs): Applications, Materials, Coating Design and Failure Mechanisms. *Metals* **2024**, *14*, 575. [CrossRef]
38. Yang, J.; Qian, X.; Pan, W.; Yang, R.; Li, Z.; Han, Y.; Zhao, M.; Huang, M.; Wan, C. Diffused Lattice Vibration and Ultralow Thermal Conductivity in the Binary Ln–Nb–O Oxide System. *Adv. Mater.* **2019**, *31*, e1808222. [CrossRef]
39. Pan, Y.; Han, D.; Huang, S.; Niu, Y.; Liang, B.; Zheng, X. Thermal insulation performance and thermal shock resistance of plasma-sprayed TiAlCrY/$Gd_2Zr_2O_7$ thermal barrier coating on γ-TiAl alloy. *Surf. Coat. Technol.* **2023**, *468*, 129715. [CrossRef]
40. Xue, Y.; Zhang, H.; Zhao, X.; An, Y.; Liu, G.; Ma, Y.; Li, H.; Zhou, H.; Chen, J. Enhanced thermal shock resistance of gradient high-entropy $(La_{0.2}Nd_{0.2}Sm_{0.2}Eu_{0.2}Gd_{0.2})_2Ce_2O_7$/YSZ thermal barrier coatings. *Ceram. Int.* **2024**, *50*, 18024–18034. [CrossRef]
41. Li, X.; Peng, X.-Y.; Dong, H.; Zhou, Y.; Wang, T.; Ren, K.; Sun, L. The Evaluation of Durability of Plasma-Sprayed Thermal Barrier Coatings with Double-layer Bond Coat. *Coatings* **2019**, *9*, 241. [CrossRef]
42. Liu, Q.; Hu, X.P.; Zhu, W.; Liu, G.L.; Guo, J.W.; Bin, J. Thermal shock performance and failure behavior of $Zr_6Ta_2O_{17}$-8YSZ double-ceramic-layer thermal barrier coatings prepared by atmospheric plasma spraying. *Ceram. Int.* **2022**, *48*, 24402–24410. [CrossRef]
43. Xu, R.; Fan, X.L.; Zhang, W.X.; Song, Y.; Wang, T.J. Effects of geometrical and material parameters of top and bond coats on the interfacial fracture in thermal barrier coating system. *Mater. Des.* **2013**, *47*, 566–574. [CrossRef]
44. Rangaraj, S.; Kokini, K. Multiple Surface Cracking and Its Effect on Interface Cracks in Functionally Graded Thermal Barrier Coatings Under Thermal Shock. *J. Appl. Mech. Mar.* **2003**, *70*, 234–245. [CrossRef]
45. Zhou, B.; Kokini, K. Effect of pre-existing surface crack morphology on the interfacial thermal fracture of thermal barrier coatings: A numerical study. *Mater. Sci. Eng. A* **2003**, *348*, 271–279. [CrossRef]

MDPI AG
Grosspeteranlage 5
4052 Basel
Switzerland
Tel.: +41 61 683 77 34

Nanomaterials Editorial Office
E-mail: nanomaterials@mdpi.com
www.mdpi.com/journal/nanomaterials

www.ingramcontent.com/pod-product-compliance
Lightning Source LLC
Chambersburg PA
CBHW062039220926
51850CB00010B/2607

* 9 7 8 3 7 2 5 8 5 8 2 9 3 *